Invertebrate Zoology

Invertebrate Zoology

Rose Watson

SYRAWOOD
PUBLISHING HOUSE
New York

Published by Syrawood Publishing House,
750 Third Avenue, 9th Floor,
New York, NY 10017, USA
www.syrawoodpublishinghouse.com

Invertebrate Zoology
Rose Watson

International Standard Book Number: 978-1-64740-033-0 (Hardback)

Cataloging-in-Publication Data

Invertebrate zoology / Rose Watson.
p. cm.
Includes bibliographical references and index.
ISBN 978-1-64740-033-0
1. Invertebrates. 2. Zoology. 3. Animals. I. Watson, Rose.
QL362 .I58 2020
592--dc23

Printed in China.

Table of Contents

Preface

It is with great pleasure that I present this book. It has been carefully written after numerous discussions with my peers and other practitioners of the field. I would like to take this opportunity to thank my family and friends who have been extremely supporting at every step in my life.

The sub-discipline of zoology, which is concerned with the study of invertebrates is referred to as invertebrate zoology. Invertebrates are the animals that do not have a backbone. It is a diverse group of animals which includes tunicates, echinoderms, sponges, various phyla of worms, arthropods, molluscs, etc. There can be many sub-divisions of invertebrate zoology including arthopodology which includes arachnology, entomology, carcinology and myriapodology. Some of the other sub-divisions of invertebrate zoology are invertebrate paleontology, malacology, and helminthology. Invertebrate paleontology deals with the scientific study of prehistoric invertebrates by analyzing invertebrate fossils in the geological records. The topics included in this book on invertebrate zoology are of utmost significance and bound to provide incredible insights to readers. Some of the diverse topics covered in this book address the varied branches that fall under this category. This book will serve as a reference to a broad spectrum of readers.

The chapters below are organized to facilitate a comprehensive understanding of the subject:

Chapter – What is Invertebrate Zoology?

Living organisms that do not possess neither develop a vertebral column are referred to as in-vertebrates. The sub-discipline of zoology that studies invertebrates is termed as invertebrate zoology. This is an introductory chapter which will briefly introduce all the significant aspects of invertebrates and invertebrate zoology.

Chapter – Branches of Invertebrate Zoology

Invertebrate zoology is an umbrella subject that comprises of different branches such as arthropodology, helminthology, malacology and paleontology. These are further divided into arachnology, entomology, carcinology, conchology, etc. The aim of this chapter is to explore these branches and sub-branches of invertebrate zoology.

Chapter – Anatomy of Invertebrates

Anatomy is the branch of biology that deals with the study of the structures of the living organisms. Porifera, cnidaria, echinoderm, platyhelminthes, nematodes, annelida, arthropoda, mollusca, etc. are some of the invertebrates studied within this field. This chapter delves into anatomy of these invertebrates to provide an extensive understanding of this discipline.

Chapter – Reproduction in Invertebrates

Reproduction is a biological procedure of producing a new individual from their parents. It is classified into sexual and asexual reproduction. In order to completely understand invertebrates, it is necessary to understand their reproduction processes. The following chapter elucidates the reproduction of invertebrates.

Chapter – Habitats of Invertebrates

The environment in which organisms live is referred to as habitat. It plays a major role in their life-cycle. Habitat is characterized by physical factors like moisture, soil, temperature, light, etc., and biotic factors like food and predators. This chapter sheds light on the habitat of invertebrates living in pond, marine and terrestrial habitats.

Rose Watson

1

What is Invertebrate Zoology?

Living organisms that do not possess neither develop a vertebral column are referred to as invertebrates. The sub-discipline of zoology that studies invertebrates is termed as invertebrate zoology. This is an introductory chapter which will briefly introduce all the significant aspects of invertebrates and invertebrate zoology.

Invertebrates

Invertebrates are animals that don't have a backbone. The vertebral column is another name for the backbone. Over 90% of all species on Earth are invertebrates, and invertebrate species have been found in the fossil record as far back as 600 million years ago. Molecular biology studies suggest that all invertebrates evolved from a single invertebrate group.

In addition to not having a backbone, invertebrates have soft bodies because they don't have an internal skeleton (endoskeleton) for support. Instead, many have structures on the outside (exoskeleton) that provide support and protection. In addition, invertebrates are cold-blooded, meaning they can't regulate their body temperature, so it changes depending on the environment.

Invertebrates are incredibly diverse. They live in fresh water, salt water, on land and as parasites in other animals. There are invertebrates that are carnivorous (meat eaters), herbivores (plant eaters) and omnivores (meat and plant eaters). There are even some invertebrate species that grow bacteria and cells inside their bodies that make their food. Some invertebrates stay in one spot, while others fly, swim, float, crawl and burrow.

Types of Invertebrates

Eighty-five percent of invertebrates – some 923,000 species – are arthropods. Mollusks have approximately 100,000 distinct species. Some of the most common types of invertebrates are:

- Protozoans – Single-celled organisms such as amoebas and paramecia.
- Annelids – Earthworms, leeches.
- Echinoderms – Starfish, sea urchins, sea cucumbers.

- Mollusks – Snails, octopi, squid, snails, clams.
- Arthropods – Insects, spiders, crustaceans such as shrimp, crabs, lobsters.

Example of Invertebrate Animals

Amoebas

Amoebas are single-celled organisms that are part of the simplest group of invertebrates. They have a cell membrane, DNA, a nucleus and organelles just like most cells in the human body. The difference is that amoebas can move using their pseudopodia, or "false feet." They also use pseudopodia to catch and eat food. Beneficial amoebas are found in soil where they help regulate bacterial populations and recycle nutrients. There are several species of ameba that cause disease, including Entameba hystolitica, which causes diarrhea.

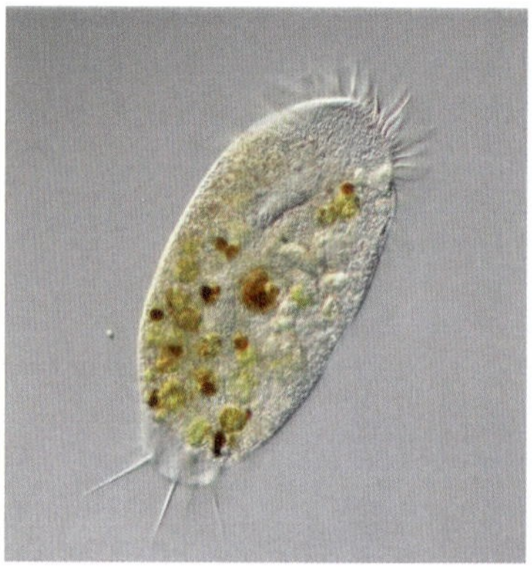

The image above shows what an ameba looks like under a microscope.

Earthworms

Earthworms, also called angleworms, are very important in the world because their burrowing turns over and aerates the soil, provides soil drainage and mixes in organic material. They are one of more than 1,800 species of worms that live on land. Earthworms eat pieces of plants and animals in the soil. The bodies of worms are divided into segments; in certain species, some of the organs are duplicated in each segment. Earthworms can detect light and vibrations, but they are blind and cannot hear. Birds and other animals use earthworms as a food source.

Starfish

The image above shows a starfish.

Starfish are also called sea stars because of the arms or "rays" they have sticking out of their bodies. Some starfish live deep in the oceans, while others live on the shore. There are over 1,600 species of starfish. Most have five arms, but one species is known to have 24 arms. Starfish have tube feet that work with hydraulic pressure to help them move. Snails, mussels and clams are common foods for starfish, which use their arms to guide food particles from the water into their mouths. These organisms have the unique ability to lose and regrow their arms.

Squids

Squids are decapods, meaning they have 10 arms or tentacles. They are different from octopuses, which have eight arms. Each tentacle has four rows of suckers that help them catch prey (usually fish) and attach themselves to surfaces. Squids live in the ocean, both near the shore and in the ocean depths. Their length ranges from about 3/4 of an inch to more than 65 feet. Some squids are luminous, which means they have light organs in their skin that let them change color to camouflage themselves, attract prey and communicate with other squids. Humans, some fishes and sperm whales eat squid.

The image above shows a reef squid.

Spiders

Spiders make their home on every continent except Antarctica. They are not insects because they have eight legs (insects have six legs). Experts think there are about 170,000 species of spiders in the world, but only around 39,000 are described and named. Some spiders live on land and others live in the water. The leg span of species ranges from 0.02 of an inch to 10 inches. Most spiders have eight eyes, and can sense light and dark. They also are very sensitive to vibration, which comes in handy when prey gets stuck in their web. Some spiders inject poison into their prey to kill it, while others prefer to wrap up their live catch and save it for later. Spiders can only ingest liquids; they bite holes in their prey and inject enzymes that dissolve the tissues so they can be sucked out.

Invertebrate Zoology

Invertebrate zoology is the scientific study of the group of animals known as invertebrates. Invertebrates are 'animals without a backbone'. Directly translated, the term means without (in-) vertebrae (-vertebrata).

A Swedish botanist Carolus Linneaus is often credited as the 'father' of modern classification as he is the one who came up with the binomial scientific naming system still used for most organismic groups today. The binomial system gives each distinct 'species' a separate first and last name: Genus species. An example would be our human scientific name, Homo sapiens. By convention, this binomial is always given with the genus capitalized, the species lower case, and the entire set either italicized OR underlined. In this way, each recognized organism has it's own unique scientific name. This uniqueness and universality across languages reduces confusion.

A species is then grouped with others of similar characteristics into the broader category known as Family (for example, humans are in the Family Hominidae, note the use of capital letters). This continues in an ever broadening nested grouping system through Order, Class, and Phylum (for our species those are Order Primate, Class Mammalia, and Phylum Chordata). And finally, we are part of the Kingdom Animalia.

The characteristics used by zoologists to group species into these various nested categories come from a variety of sources. Historically, external (outside) morphology and internal(inside) anatomy were used most often, together with details about life cycle and development (i.e. steps between fertilization, through larval stages, and into patterning of the adult body). In modern times, comparison of DNA sequences (molecular genetics) as well as aspects of regulation within and between cells (physiology, cell biology, and developmental genetics) have been added to the process of classification (taxonomy). In this way, taxonomy relies on a variety of character sets to build the nested categories of classification.

2

Branches of Invertebrate Zoology

Invertebrate zoology is an umbrella subject that comprises of different branches such as arthropodology, helminthology, malacology and paleontology. These are further divided into arachnology, entomology, carcinology, conchology, etc. The aim of this chapter is to explore these branches and sub-branches of invertebrate zoology.

Arthropodology

Arthropodology is a biological discipline that focuses on the arthropods. Arthropods are invertebrates that belong to the phylum Arthropoda. This phylum is the largest among the phyla in the animal kingdom. They include the insects, the arachnids, the myriapods and the crustaceans. These animals are grouped under this phylum based on the features common to them. They have multiple paired jointed limbs and a hard chitinous exoskeleton. The exoskeleton is a hard outer casing that protects their soft internal organs. Arthropodology deals with these animals and aims to understand their evolution, anatomy, biology, and ecological importance.

Arthropodology that studies the parasitic mode of life of certain arthropods is referred to as medical arthropodology. Apart from parasitic arthropods, there are also arthropod species that act as vectors of disease. An example of that would be the mosquito. Certain species of mosquitoes serve as vectors for disease-causing viruses and parasites.

Mosquito-borne viral diseases include the yellow fever, dengue fever, and zika. Mosquito-borne parasitic diseases are malaria and lymphatic filariasis.

Arachnology

Arachnology is the scientific study of spiders and related animals such as scorpions, pseudoscorpions, and harvestmen, collectively called arachnids. Those who study spiders and other arachnids are arachnologists. More narrowly, the study of spiders alone (order Araneae) is known as araneology.

Nephila clavipes.

Arachnology as a Science

Arachnologists are primarily responsible for classifying arachnids and studying aspects of their biology. In the popular imagination, they are sometimes referred to as spider experts. Disciplines within arachnology include naming species and determining their evolutionary relationships to one another (taxonomy and systematics), studying how they interact with other members of their species and/or their environment (behavioural ecology), or how they are distributed in different regions and habitats (faunistics). Other arachnologists perform research on the anatomy or physiology of arachnids, including the venom of spiders and scorpions. Others study the impact of spiders in agricultural ecosystems and whether they can be used as biological control agents.

Subdisciplines

Arachnology can be broken down into several specialties. These topics include:

- Acarology – The study of ticks and mites.
- Araneology – The study of spiders.
- Scorpiology – The study of scorpions.

Acarology

Acarology is the study of mites and ticks, the animals in the order Acarina. It is a subfield of arachnology, a subdiscipline of the field of zoology. A zoologist specializing in acarology is called an acarologist.

Ticks

Ticks are arachnids, typically 3 to 5 mm long, part of the order Parasitiformes. Along with mites, they constitute the subclass Acari. Ticks are external parasites, living by feeding on the blood of mammals, birds, and sometimes reptiles and amphibians. Ticks evolved by the Cretaceous period, the most common form of fossilisation being amber immersion. Ticks are widely distributed around the world, especially in warm, humid climates.

Almost all ticks belong to one of two major families, the Ixodidae or hard ticks, and the Argasidae or soft ticks. Adults have ovoid or pear-shaped bodies, which become engorged with blood when they feed, and eight legs. In addition to having a hard shield on their dorsal surfaces, hard ticks have a beak-like structure at the front containing the mouthparts, whereas soft ticks have their mouthparts on the underside of their bodies. Both families locate a potential host by odour or from changes in the environment.

Ticks have four stages to their lifecycle, namely egg, larva, nymph, and adult. Ixodid ticks have three hosts, taking at least a year to complete their lifecycle. Argasid ticks have up to seven nymphal stages (instars), each one requiring a blood meal. Because of their habit of ingesting blood, ticks are vectors of many diseases that affect humans and other animals.

Biology

Taxonomy and Phylogeny

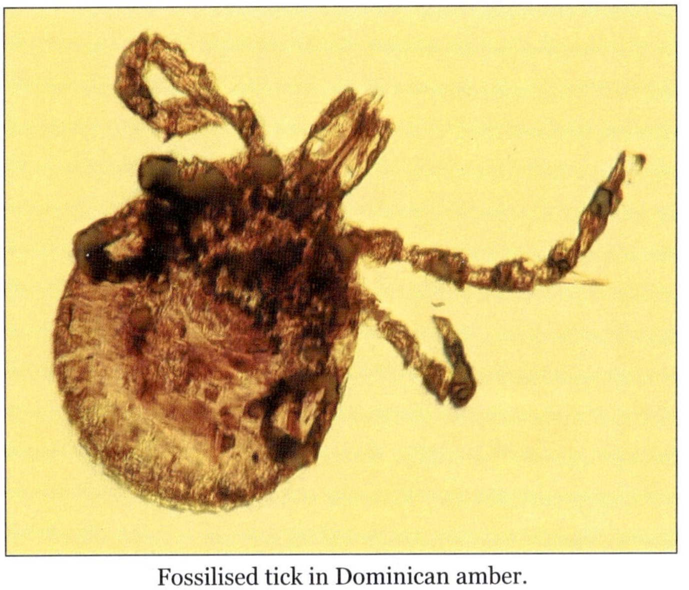

Fossilised tick in Dominican amber.

Fossilized ticks are known from the Cretaceous onwards, most commonly in amber. They most likely originated in the Cretaceous (146 to 66 million years ago), with most of the evolution and dispersal occurring during the Tertiary (65 to 5 million years ago). The oldest example is an argasid bird tick from Cretaceous New Jersey amber. The younger Baltic and Dominican ambers have also yielded examples that can be placed in living genera. The tick *Deinocroton draculi* has been found with dinosaur feathers preserved in Cretaceous Burmese amber from 99 million years ago.

Three families of ticks are described. The two large ones are the sister families of Ixodidae (hard ticks) and Argasidae (soft ticks). The third is the Nuttalliellidae, named for the bacteriologist George Nuttall. It comprises a single species, *Nuttalliella namaqua*, and is the most basal lineage. It is found in southern Africa from Tanzania to Namibia and South Africa. Ticks are closely related to the mites, within the subclass Acari. rDNA analysis suggests that the Ixodidae are a clade, but that the Argasidae may be paraphyletic.

The Ixodidae contain over 700 species of hard ticks with a scutum or hard shield, which the Argasidae lack. The Argasidae contain about 200 species; the genera accepted as of 2010 are *Antricola*, *Argas*, *Nothoaspis*, *Ornithodoros*, and *Otobius*. They have no scutum, and the capitulum (mouth and feeding parts) is concealed beneath the body.The phylogeny of the Ixodida within the Acari is shown in the cladogram, based on a 2014 maximum parsimony study of amino acid sequences of 12 mitochondrial proteins. The Argasidae appear monophyletic in this study.

Range and Habitat

Tick species are widely distributed around the world, but they tend to flourish more in countries with warm, humid climates, because they require a certain amount of moisture in the air to undergo metamorphosis, and because low temperatures inhibit their development from eggs to larvae. Ticks are also widely distributed among host taxa, which include marsupial and placental mammals, birds, reptiles (snakes, iguanas, and lizards), and amphibians. Ticks of domestic animals cause considerable harm to livestock by transmission of many species of pathogens, as well as causing anaemia and damaging wool and hides. Some of the most debilitating species occur in tropical countries. Tropical bont ticks affect most domestic animals and occur in Africa and the Caribbean. The spinose ear tick has a worldwide distribution, the young feeding inside the ears of cattle and wild animals.

In general, ticks are to be found wherever their host species occur. Migrating birds carry ticks with them on their journeys; a study of migratory birds passing through Egypt found more than half the bird species examined were carrying ticks. The species of tick often differed between the autumn and spring migrations, probably because of the seasonal periodicities of the different species.

For an ecosystem to support ticks, it must satisfy two requirements; the population density of host species in the area must be high enough, and humidity must be high enough for ticks to remain hydrated. Due to their role in transmitting Lyme disease, ixodid ticks, particularly the North American *I. scapularis*, have been studied using geographic information systems to develop predictive models for ideal tick habitats. According to these studies, certain features of a given microclimate – such as sandy soil, hardwood trees, rivers, and the presence of deer – were determined to be good predictors of dense tick populations.

A habitat preferred by ticks is the interface where a lawn meets the woods, or more generally, the ecotone, which is unmaintained transitional edge habitat between woodlands and open areas. Therefore, one tick management strategy is to remove leaf litter, brush, and weeds at the edge of the woods. Ticks like shady, moist leaf litter with an overstory of trees or at least shrubs, and they deposit their eggs into such places in the spring, so that the larvae can emerge in the fall and crawl into low-lying vegetation. The 3 m of boundary closest to the lawn's edge are a tick migration zone, where 82% of tick nymphs in lawns are found.

Anatomy and Physiology

A hard-bodied tick of the family Ixodidae, the lone star tick.

Ticks, like mites, are arthropods that have lost the segmentation of the abdomen that their ancestors had, with a subsequent fusion of the abdomen with the cephalothorax. The tagmata typical of other Chelicerata have been replaced by two new body sections, the anterior capitulum (or gnathosoma), which is retractable and contains the mouthparts, and the posterior idiosoma, which contains the legs, digestive tract, and reproductive organs. The capitulum is a feeding structure with mouthparts adapted for piercing skin and sucking blood; it is only the front of the head and contains neither the brain nor the eyes. Features of the capitulum include the basis capitulum, two palps, two cutting chelicerae, and hypostome. The basis capitulum supports the rest of the feeding structures. Palps have a sensory role and are composed of three sections. The hypostome is used for blood extraction and is a hollow, tube-like structure.

The ventral side of the idiosoma bears sclerites, and the gonopore is located between the fourth pair of legs. In the absence of segmentation, the positioning of the eyes, limbs, and gonopore on the idiosoma provide the only locational guidance.

Most ticks are inornate, meaning they are reddish or mahogany in color and lack markings. However, some species are ornate and have distinctive white patterns on the scutum.

Larval ticks hatch with six legs, acquiring the other two after a blood meal and molting into the nymph stage. In the nymphal and adult stages, ticks have eight legs, each of which has seven segments and is tipped with a pair of claws. The legs are sometimes ornamented and usually bear sensory or tactile hairs. In addition to being used for locomotion, the tarsus of leg I contains a unique sensory structure, Haller's organ, which can detect odors and chemicals emanating from the host, as well as sensing changes in temperature and air currents. Ticks can also use Haller's organs to perceive infrared light emanating from a host. When not being used for walking, the legs remain tightly folded against the body.

In nymphs and adults of the family Ixodidae, the *capitulum* is prominent and projects forwards from the body, a feature not present in the Argasidae. The eyes are close to the sides of the scutum, and the large spiracles are located just behind the coxae of the fourth pair of legs.

The hard, protective scutellum, characteristic of this family, covers the whole dorsal surface in males, but is restricted to a small, shield-like structure behind the capitulum in females and nymphs. They differ, too, in their lifecycle; Ixodidae that attach to a host can bite painlessly and generally are unnoticed, and they remain in place until they engorge and are ready to change their skin; this process may take days or weeks. Some species drop off the host to moult in a safe place, whereas others remain on the same host and only drop off once they are ready to lay their eggs.

A soft-bodied tick of the family Argasidae, beside eggs it has just laid.

The body of the soft tick, family Argasidae, is pear-shaped or oval with a rounded anterior portion. The mouthparts cannot be seen from above, as they are on the ventral surface. The cuticle is leathery; often, a centrally positioned dorsal plate is seen, with ridges that project slightly above the surrounding surface, but with no decoration. A pattern of small, circular, depressed areas shows where muscles are attached to the interior of the integument. The eyes are on the sides of the body, the spiracles open between legs 3 and 4, and males and females only differ in the structure of the genital pore.

The Nuttalliellidae can be distinguished from both ixodid and argasid ticks by a combination of a projecting capitulum at the front and a soft, leathery skin. Other distinguish characteristics include the position of the stigmata, the lack of setae, the strongly corrugated integument, and the form of the fenestrated plates.

Ticks are extremely tough, hardy, and resilient animals. They can survive in a near-vacuum for as long as a half hour. Their slow metabolism during their dormant periods enables them to go long periods between meals. During droughts, they can endure dehydration without feeding for as long as 18 weeks, although ticks with limited energy reserves may succumb to desiccation after 36 weeks. To keep from dehydrating, ticks hide in humid spots on the forest floor or absorb water from subsaturated air by secreting hygroscopic fluid produced by the salivary glands onto the external mouthparts and then reingesting the water-enriched fluid.

Ticks can withstand temperatures just above 0 °F (−18 °C) for more than two hours, and can survive temperatures in the 20–29 °F (−7 - −2 °C) range for at least two weeks. Ticks have even been found in Antarctica, where they feed on penguins.

Diet and Feeding

A questing tick, fingers for scale.

Ticks satisfy all of their nutritional requirements as ectoparasites, feeding on a diet of blood. They are obligate hematophages, needing blood to survive and move from one stage of life to another. Ticks can fast for long periods, but eventually die if unable to find a host. This behavior evolved approximately 120 million years ago through adaptation to blood-feeding. The behavior evolved independently in the separate tick families, with differing host-tick interactions driving the evolutionary change.

Some ticks attach quickly, while others wander around looking for thinner skin, such as is found on the ears of mammals. Depending on the species and life stage, preparing to feed can take from 10 minutes to two hours. On locating a suitable feeding spot, the tick grasps the host's skin and cuts into the surface. They extract blood by cutting a hole in the host's epidermis, into which they insert their hypostome, and keep the blood from clotting by excreting an anticoagulant or platelet aggregation inhibitor.

Ticks find their hosts by detecting animals' breath and body odors, or by sensing body heat, moisture, and vibrations. They are incapable of flying or jumping, but many tick species, particularly Ixodidae, lie in wait in a position known as "questing". While questing, ticks cling to leaves and grasses by their third and fourth pairs of legs. They hold the first pair of legs outstretched, waiting to grasp and climb on to any passing host. Tick questing heights tend to be correlated with the size of the desired host; nymphs and small species tend to quest close to the ground, where they may encounter small mammalian or bird hosts; adults climb higher into the vegetation, where larger hosts may be encountered. Some species are hunters and lurk near places where hosts may rest. On receiving an olfactory or other stimulus, they crawl or run across the intervening surface.

Other ticks, mainly the Argasidae, are "nidicolous", finding hosts in their nests or burrows, and in caves in the case of bats. They use the same stimuli as non-nidicolous species to identify hosts, with body heat and odors often being the main factors. Many of them feed primarily on birds, though some *Ornithodoros* species, for example, feed on mammals. Both groups feed rapidly, typically biting painfully and drinking their fill within minutes. None of the species sticks to the host in the way that hard ticks do. Unlike the Ixodidae that have no fixed dwelling place except on the host, they live in sand or in crevices near animal dens or nests, or in human dwellings where they come out nightly to attack roosting birds, or emerge when they detect carbon dioxide in the breath of their hosts.

In the Ixodidae, the tick stays in place until it is completely engorged. Its weight may increase by 200 to 600 times as compared to its weight before it started feeding. To accommodate this large expansion, cell division takes place and its cuticle grows larger; the tick may remain attached for days or weeks, depending on species, life stage, and host. In the Argasidae, the tick's cuticle stretches to accommodate the fluid ingested, but does not grow new cells, with the weight of the tick increasing five- to 10-fold over the unfed state. The tick then drops off the host and typically remains in the nest or burrow until its host returns to provide its next meal.

Tick saliva contains about 1,500 to 3,000 proteins depending on the tick species. The proteins with anti-inflammatory properties, called evasins, help ticks to feed for 8–10 days without being noticed by the host animal, as they block the host's chemokines and prevent painful inflammation. Researchers are studying these evasins with the goal of developing drugs to neutralise the chemokines that cause inflammation in myocarditis, heart attack, and stroke.

Ecology

Mites and nematodes feed on ticks, which are also a minor nutritional resource for birds. More importantly, they carry diseases as the primary hosts of pathogens such as spirochaetes, and without their agency, the organisms could not reach their secondary hosts. The diseases caused may debilitate their victims, and ticks may thus be assisting in controlling animal populations and preventing overgrazing.

Certain infectious diseases of humans and other animals can be transmitted by ticks, with the species of tick involved tending to be those with a wide host range. Spread of disease in this way is enhanced by the extended time during which a tick remains attached, during which time the mobile host can be carried long distances, or in the case of bird hosts, across the sea. The infective agents can be present not only in the adult tick, but also in the eggs produced plentifully by the females. Many tick species have extended their ranges as a result of the movements of people, their pets, and livestock. With increasing participation in outdoor activities such as wilderness hikes, more people and their dogs may find themselves exposed to attack.

Life Cycle

Both ixodid and argasid ticks have four lifecycle stages: egg, larva, nymph, and adult.

One-host Ticks

Eggs laid in the environment hatch into larvae, which immediately seek out a host in which to attach and feed. Fed larvae molt into unfed nymphs that remain on the host. After engorging on the host's blood, the nymphs molt into sexually mature adults that remain on the host in order to feed and mate. Once a female is both fed and ready to lay eggs, only then does she drop off the host in search of a suitable area to deposit her eggs. Ticks that follow this life cycle are called one-host ticks. The winter tick *Dermacentor albipictus* and the cattle tick *Boophilus microplus* are examples of one-host ticks.

Two-host Ticks

Newly hatched larvae attach to a host in order to obtain a blood meal. They remain on the host after developing into nymphs. After emerging from their shed larval skins, the nymphs reattach to the host and feed. Once engorged, they drop off the host and find a safe area in the natural environment in which to molt into adults. Both male and female adults seek out a host on which to attach, which may be the same body that served as host during their early development. Once attached, they feed and mate. Females ready to lay eggs drop from the host to oviposit in the environment. Ticks that complete their life cycle in this manner are called two-host ticks, like *Hyalomma anatolicum excavatum*.

Three-host Ticks

Most ixodid ticks require three hosts, and their lifecycles take at least a year to complete. Thousands of eggs are laid on the ground by an adult female tick. When the larvae emerge, they attach and feed primarily on small mammals and birds. After feeding, they detach from their hosts and molt to nymphs on the ground, which then attach and feed on larger hosts before dropping off yet again in order to molt into adults. Adults seek out a third host on which to feed and mate. Female adults engorge on blood and prepare to drop off to lay her eggs on the ground, while males feed very little and remain on the host in order to continue mating with other females.

Argasidae

Argasid ticks, unlike ixodid ticks, may go through up to seven nymphal stages (instars), requiring a meal of blood each time. Their lifecycles range from months to years. The adult female argasid tick can lay a few hundred to over a thousand eggs over the course of her lifetime. Larvae feed very quickly and detach to molt into nymphs. Both male and female adults feed on blood, and they mate off the host. During feeding, any excess fluid is excreted by the coxal glands, a process that is unique to argasid ticks.

Relationship with Humans

Tick-borne Disease

Ticks are implicated in the transmission of a number of infections caused by pathogens such as bacteria, viruses, and protozoa. Sometimes, the tick harbours more than one type of pathogen, making diagnosis of the infection more difficult. Species of the bacterial genus *Rickettsia* are responsible for typhus, rickettsialpox, boutonneuse fever, African tick bite fever, Rocky Mountain spotted fever, Flinders Island spotted fever, and Queensland tick typhus (Australian tick typhus). Other tick-borne diseases include Lyme disease and Q fever, Colorado tick fever, Crimean Congo hemorrhagic fever, tularemia, tick-borne relapsing fever, babesiosis, ehrlichiosis, Bourbon virus, and tick-borne meningoencephalitis, as well as bovine anaplasmosis and probably the Heartland virus.

Some species, notably the Australian paralysis tick, are also intrinsically venomous and can cause tick paralysis. Eggs can become infected with pathogens inside a female tick's ovaries, in which case the larval ticks are infectious immediately at hatching, before feeding on their first host. Tropical bont ticks transmit the rickettsial disease heartwater, which can be particularly devastating in cattle. The ticks carried by migratory birds may act as reservoirs and vectors of infectious diseases. Over 20 strains of pathogenic viruses were found in the autumn in the Egyptian migratory bird study.

Not all ticks in an infective area are infected with pathogens, and both attachment of the tick and a long feeding session seem to be necessary for transmission of these diseases to take place. Thus, tick bites often do not lead to infection, especially if the ticks are removed within 36 hours. Adult ticks can be removed with fine-tipped tweezers or proprietary tick removal tools, disinfecting the wound. It is also possible to freeze them off with a medical wart remover. If the tick's head and mouthparts break off during removal, they can be removed with tweezers like a splinter. Ticks can be disposed of by flushing them down the toilet, placing them in a container of soapy water or alcohol, or sticking them to tape that can then be folded over and thrown away.

Bifenthrin and permethrin, both pyrethroids, are sometimes used as tick-control measures, although a downside is that they are carcinogenic and can attack the nervous systems of other species besides ticks. Those who walk through tick-infested areas can make it harder for ticks to latch onto them by tucking their slacks into boots made of smooth rubber, which ticks have more trouble climbing than other material.

Research since 2008 has documented red-meat allergies (known as Alpha-gal syndrome) in the U.S. due to lone star tick bites. The range of the problem has been expanding with the range of the tick. Other species of ticks are suspected of being responsible for meat allergies in other countries, including Sweden, Germany, and Australia.

Many tick-transmitted viruses, such as Crimean-Congo hemorrhagic fever virus, Kyasanur Forest disease virus, Alkhumra hemorrhagic fever virus, and Omsk hemorrhagic fever virus, are classified as dangerous enough to require biosafety level 4 precautions in laboratory environments. This includes five levels of containment, viz., storage vials within humidified desiccators, within environmental chambers, within a tick suite, within a BSL4 laboratory. Precautions such as glove boxes, sticky pads, Vaseline barriers, safety suits, gloves, sticky tape, silicone vacuum grease, sticky trap paste, and micro mesh are used to safely contain ticks and prevent them from escaping.

A sign in a Lithuanian forest warning of high risk of tick-borne encephalitis infection.

Population Control Measures

Researcher collecting ticks using the "tick dragging" method.

With the possible exception of widespread DDT use in the Soviet Union, attempts to limit the population or distribution of disease-causing ticks have been quite unsuccessful. The parasitoid chalcid wasp *Ixodiphagus hookeri* has been investigated for its potential to control tick populations. It lays its eggs into ticks; the hatching wasps kill their hosts.

Predators and competitors of tick hosts can indirectly reduce the density of infected nymphs, thereby lowering tick-borne disease risk by lowering the density and/or tick burden of reservoir-competent hosts. A study in the Netherlands found that the number of larval ticks on bank voles and wood mice was lower at sites with significant red fox (*Vulpes vulpes*) and stone marten (*Martes foina*) activity.

This supports the results of a study from the northeastern US, in which the incidence of Lyme borreliosis was negatively correlated with the density of red fox, which was suggested to be because foxes decrease the density of white-footed mice (*Peromyscus leucopus*), the most important reservoir-competent host for *Borrelia burgdorferi* in the US.

Another natural form of control for ticks is the guineafowl, a bird species that consumes mass quantities of ticks. Opossums also are net destroyers of ticks, killing around 90% of the ticks that attempt to attach and feed on them. More generally, high animal diversity has a strongly protective effect against tick-borne disease.

Topical tick medicines may be toxic to animals and humans. The synthetic pyrethroid insecticide phenothrin in combination with the hormone analogue methoprene was a popular topical flea and tick therapy for felines. Phenothrin kills adult ticks, while methoprene kills eggs. However, some products have been withdrawn, while others are known to cause adverse reactions.

Araneology

Araneology is the study of spiders, the animals in the class Arachnida, order Araneae. It is a subfield of arachnology, a subdiscipline of the field of zoology. A zoologist specializing in araneology is called an Araneologist.

Spiders

Spiders (order Araneae) are air-breathing arthropods that have eight legs and chelicerae with fangs able to inject venom. They are the largest order of arachnids and rank seventh in total species diversity among all orders of organisms. Spiders are found worldwide on every continent except for Antarctica, and have become established in nearly every habitat with the exceptions of air and sea colonization. As of July 2019, at least 48,200 spider species, and 120 families have been recorded by taxonomists. However, there has been dissension within the scientific community as to how all these families should be classified, as evidenced by the over 20 different classifications that have been proposed since 1900.

Anatomically, spiders differ from other arthropods in that the usual body segments are fused into two tagmata, the cephalothorax and abdomen, and joined by a small, cylindrical pedicel. Unlike insects, spiders do not have antennae. In all except the most primitive group, the Mesothelae, spiders have the most centralized nervous systems of all arthropods, as all their ganglia are fused into one mass in the cephalothorax. Unlike most arthropods, spiders have no extensor muscles in their limbs and instead extend them by hydraulic pressure.

Their abdomens bear appendages that have been modified into spinnerets that extrude silk from up to six types of glands. Spider webs vary widely in size, shape and the amount of sticky thread used. It now appears that the spiral orb web may be one of the earliest forms, and spiders that produce tangled cobwebs are more abundant and diverse than orb-web spiders. Spider-like arachnids with silk-producing spigots appeared in the Devonian period about 386 million years ago, but these animals apparently lacked spinnerets. True spiders have been found in Carboniferous rocks from 318 to 299 million years ago, and are very similar to the most primitive surviving suborder, the Mesothelae. The main groups of modern spiders, Mygalomorphae and Araneomorphae, first appeared in the Triassic period, before 200 million years ago.

The species *Bagheera kiplingi* was described as herbivorous in 2008, but all other known species are predators, mostly preying on insects and on other spiders, although a few large species also take birds and lizards. It is estimated that the world's 25 million tons of spiders kill 400–800 million tons of prey per year. Spiders use a wide range of strategies to capture prey: trapping it in sticky webs, lassoing it with sticky bolas, mimicking the prey to avoid detection, or running it down. Most detect prey mainly by sensing vibrations, but the active hunters have acute vision, and hunters of the genus *Portia* show signs of intelligence in their choice of tactics and ability to develop new ones. Spiders' guts are too narrow to take solids, so they liquefy their food by flooding it with digestive enzymes. They also grind food with the bases of their pedipalps, as arachnids do not have the mandibles that crustaceans and insects have.

To avoid being eaten by the females, which are typically much larger, male spiders identify themselves to potential mates by a variety of complex courtship rituals. Males of most species survive a few matings, limited mainly by their short life spans. Females weave silk egg-cases, each of which may contain hundreds of eggs. Females of many species care for their young, for example by carrying them around or by sharing food with them. A minority of species are social, building communal webs that may house anywhere from a few to 50,000 individuals. Social behavior ranges from precarious toleration, as in the widow spiders, to co-operative hunting and food-sharing. Although most spiders live for at most two years, tarantulas and other mygalomorph spiders can live up to 25 years in captivity.

While the venom of a few species is dangerous to humans, scientists are now researching the use of spider venom in medicine and as non-polluting pesticides. Spider silk provides a combination of lightness, strength and elasticity that is superior to that of synthetic materials, and spider silk genes have been inserted into mammals and plants to see if these can be used as silk factories. As a result of their wide range of behaviors, spiders have become common symbols in art and mythology symbolizing various combinations of patience, cruelty and creative powers. An abnormal fear of spiders is called arachnophobia.

Body Plan

1: pedipalp
2: trichobothria
3: carapace of prosoma (cephalothorax)
4: opisthosoma (abdomen)
5: eyes – AL (anterior lateral)
AM (anterior median)
PL (posterior lateral)
PM (posterior median)
Leg segments:
6: costa
7: trochanter
8: patella
9: tibia
10: metatarsus
11: tarsus
13: claw
14: chelicera

Palystes castaneus female dorsal aspect.

Spiders are chelicerates and therefore arthropods. As arthropods they have: segmented bodies with jointed limbs, all covered in a cuticle made of chitin and proteins; heads that are composed of several segments that fuse during the development of the embryo. Being chelicerates, their bodies consist of two tagmata, sets of segments that serve similar functions: the foremost one, called the cephalothorax or prosoma, is a complete fusion of the segments that in an insect would form two separate tagmata, the head and thorax; the rear tagma is called the abdomen or opisthosoma. In spiders, the cephalothorax and abdomen are connected by a small cylindrical section, the pedicel. The pattern of segment fusion that forms chelicerates' heads is unique among arthropods, and what would normally be the first head segment disappears at an early stage of development, so that chelicerates lack the antennae typical of most arthropods. In fact, chelicerates' only appendages ahead of the mouth are a pair of chelicerae, and they lack anything that would function directly as "jaws". The first appendages behind the mouth are called pedipalps, and serve different functions within different groups of chelicerates.

Spiders and scorpions are members of one chelicerate group, the arachnids. Scorpions' chelicerae have three sections and are used in feeding. Spiders' chelicerae have two sections and terminate in fangs that are generally venomous, and fold away behind the upper sections while not in use. The upper sections generally have thick "beards" that filter solid lumps out of their food, as spiders can take only liquid food. Scorpions' pedipalps generally form large claws for capturing prey, while those of spiders are fairly small appendages whose bases also act as an extension of the mouth; in addition, those of male spiders have enlarged last sections used for sperm transfer.

In spiders, the cephalothorax and abdomen are joined by a small, cylindrical pedicel, which enables the abdomen to move independently when producing silk. The upper surface of the cephalothorax is covered by a single, convex carapace, while the underside is covered by two rather flat plates. The abdomen is soft and egg-shaped. It shows no sign of segmentation, except that the primitive Mesothelae, whose living members are the Liphistiidae, have segmented plates on the upper surface.

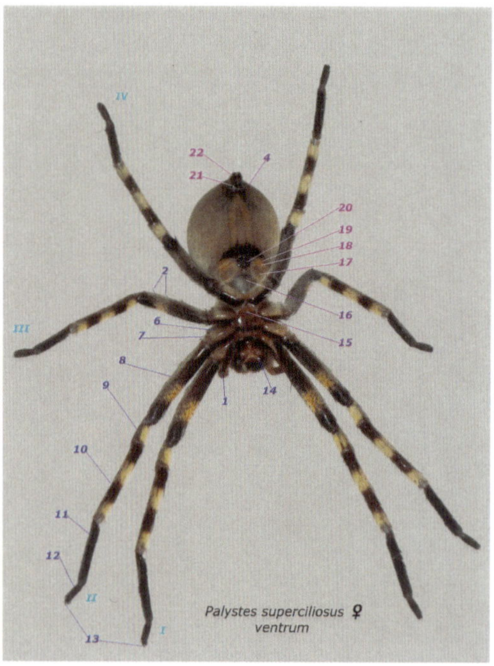

15: sternum of prosoma
16: pedicel (also called pedicle)
17: book lung sac
18: book lung stigma
19: epigastric fold
20: epigyne
21: anterior spinneret
22: posterior spinneret
I, II, III, IV = Leg numbers from anterior to posterior

Nos 1 to 14 as for dorsal aspect.

Circulation and Respiration

Like other arthropods, spiders are coelomates in which the coelom is reduced to small areas round the reproductive and excretory systems. Its place is largely taken by a hemocoel, a cavity that runs most of the length of the body and through which blood flows. The heart is a tube in the upper part of the body, with a few ostia that act as non-return valves allowing blood to enter the heart from the hemocoel but prevent it from leaving before it reaches the front end. However, in spiders, it occupies only the upper part of the abdomen, and blood is discharged into the hemocoel by one artery that opens at the rear end of the abdomen and by branching arteries that pass through the pedicle and open into several parts of the cephalothorax. Hence spiders have open circulatory systems. The blood of many spiders that have book lungs contains the respiratory pigment hemocyanin to make oxygen transport more efficient.

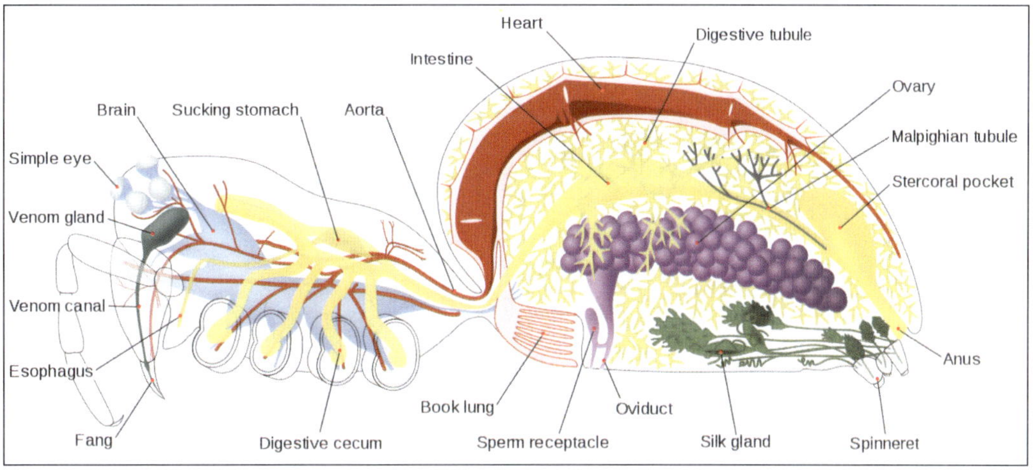

Spiders have developed several different respiratory anatomies, based on book lungs, a tracheal system, or both. Mygalomorph and Mesothelae spiders have two pairs of book lungs filled with haemolymph, where openings on the ventral surface of the abdomen allow air to enter and diffuse oxygen. This is also the case for some basal araneomorph spiders, like the family Hypochilidae, but the remaining members of this group have just the anterior pair of book lungs intact while the posterior pair of breathing organs are partly or fully modified into tracheae, through which oxygen is diffused into the haemolymph or directly to the tissue and organs. The trachea system has most likely evolved in small ancestors to help resist desiccation. The trachea were originally connected to the surroundings through a pair of openings called spiracles, but in the majority of spiders this pair of spiracles has fused into a single one in the middle, and moved backwards close to the spinnerets. Spiders that have tracheae generally have higher metabolic rates and better water conservation. Spiders are ectotherms, so environmental temperatures affect their activity.

A syrphid fly captured in the web of a spider.

Feeding, Digestion and Excretion

Cheiracanthium punctorium, displaying fangs.

Uniquely among chelicerates, the final sections of spiders' chelicerae are fangs, and the great majority of spiders can use them to inject venom into prey from venom glands in the roots of the chelicerae. The families Uloboridae and Holarchaeidae, and some Liphistiidae spiders, have lost their venom glands, and kill their prey with silk instead. Like most arachnids, including scorpions, spiders have a narrow gut that can only cope with liquid food and two sets of filters to keep solids out. They use one of two different systems of external digestion. Some pump digestive enzymes from the midgut into the prey and then suck the liquified tissues of the prey into the gut, eventually leaving behind the empty husk of the prey. Others grind the prey to pulp using the chelicerae and the bases of the pedipalps, while flooding it with enzymes; in these species, the chelicerae and the bases of the pedipalps form a preoral cavity that holds the food they are processing.

The stomach in the cephalothorax acts as a pump that sends the food deeper into the digestive system. The mid gut bears many digestive ceca, compartments with no other exit, that extract nutrients from the food; most are in the abdomen, which is dominated by the digestive system, but a few are found in the cephalothorax.

Most spiders convert nitrogenous waste products into uric acid, which can be excreted as a dry material. Malphigian tubules ("little tubes") extract these wastes from the blood in the hemocoel and dump them into the cloacal chamber, from which they are expelled through the anus. Production of uric acid and its removal via Malphigian tubules are a water-conserving feature that has evolved independently in several arthropod lineages that can live far away from water, for example the tubules of insects and arachnids develop from completely different parts of the embryo. However, a few primitive spiders, the sub-order Mesothelae and infra-order Mygalomorphae, retain the ancestral arthropod nephridia ("little kidneys"), which use large amounts of water to excrete nitrogenous waste products as ammonia.

Central Nervous System

The basic arthropod central nervous system consists of a pair of nerve cords running below the gut, with paired ganglia as local control centers in all segments; a brain formed by fusion of the ganglia for the head segments ahead of and behind the mouth, so that the esophagus is encircled by this conglomeration of ganglia. Except for the primitive Mesothelae, of which the Liphistiidae are the sole surviving family, spiders have the much more centralized nervous system that is typical of arachnids: *all* the ganglia of all segments behind the esophagus are fused, so that the cephalothorax is largely filled with nervous tissue and there are no ganglia in the abdomen; in the Mesothelae, the ganglia of the abdomen and the rear part of the cephalothorax remain unfused.

Despite the relatively small central nervous system, some spiders (like *Portia*) exhibit complex behavior, including the ability to use a trial-and-error approach.

Sense Organs

Eyes

This jumping spider's main ocelli (center pair) are very acute. The outer pair are "secondary eyes" and there are other pairs of secondary eyes on the sides and top of its head.

Spiders have primarily four pairs of eyes on the top-front area of the cephalothorax, arranged in patterns that vary from one family to another. The principal pair at the front are of the type called pigment-cup ocelli ("little eyes"), which in most arthropods are only capable of detecting the direction from which light is coming, using the shadow cast by the walls of the cup. However, in spiders these eyes are capable of forming images. The other pairs, called secondary eyes, are thought to be derived from the compound eyes of the ancestral chelicerates, but no longer have the separate facets typical of compound eyes. Unlike the principal eyes, in many spiders these secondary eyes detect light reflected from a reflective tapetum lucidum, and wolf spiders can be spotted by torch light reflected from the tapeta. On the other hand, jumping spiders' secondary eyes have no tapeta.

Other differences between the principal and secondary eyes are that the latter have rhabdomeres that point away from incoming light, just like in vertebrates, while the arrangement is the opposite in the former. The principal eyes are also the only ones with eye muscles, allowing them to move the retina. Having no muscles, the secondary eyes are immobile.

Some jumping spiders' visual acuity exceeds by a factor of ten that of dragonflies, which have by far the best vision among insects; in fact the human eye is only about five times sharper than a jumping spider's. They achieve this by a telephoto-like series of lenses, a four-layer retina and the ability to swivel their eyes and integrate images from different stages in the scan. The downside is that the scanning and integrating processes are relatively slow.

There are spiders with a reduced number of eyes. Of these, those with six eyes (such as *Periegops suterii*) are the most numerous and are missing a pair of eyes on the anterior median line; other species have four eyes and some just two. Cave dwelling species have no eyes, or possess vestigial eyes incapable of sight.

Other Senses

As with other arthropods, spiders' cuticles would block out information about the outside world, except that they are penetrated by many sensors or connections from sensors to the nervous system. In fact, spiders and other arthropods have modified their cuticles into elaborate arrays of sensors. Various touch sensors, mostly bristles called setae, respond to different levels of force, from strong contact to very weak air currents. Chemical sensors provide equivalents of taste and smell, often by means of setae. An adult *Araneus* may have up to 1000 such chemosensitive setae, most on the tarsi of the first pair of legs. Males have more chemosensitive hairs on their pedipalps than females. They have been shown to be responsive to sex pheromones produced by females, both contact and air-borne. The jumping spider *Evarcha culicivora* uses the scent of blood from mammals and other vertebrates, which is obtained by capturing blood-filled mosquitoes, to attract the opposite sex. Because they are able to tell the sexes apart, it is assumed the blood scent is mixed with pheromones. Spiders also have in the joints of their limbs slit sensillae that detects force and vibrations. In web-building spiders, all these mechanical and chemical sensors are more important than the eyes, while the eyes are most important to spiders that hunt actively.

Like most arthropods, spiders lack balance and acceleration sensors and rely on their eyes to tell them which way is up. Arthropods' proprioceptors, sensors that report the force exerted by muscles and the degree of bending in the body and joints, are well understood. On the other hand, little is known about what other internal sensors spiders or other arthropods may have.

Locomotion

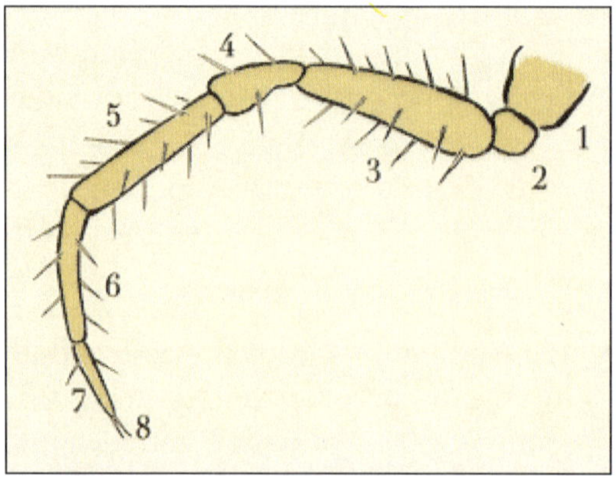

Image of a spider leg: 1–coxa; 2–trochanter; 3–femur; 4–patella; 5–tibia; 6–metatarsus; 7–tarsus; 8–claws.

Each of the eight legs of a spider consists of seven distinct parts. The part closest to and attaching the leg to the cephalothorax is the coxa; the next segment is the short trochanter that works as a hinge for the following long segment, the femur; next is the spider's knee, the patella, which acts as the hinge for the tibia; the metatarsus is next, and it connects the tibia to the tarsus (which may be thought of as a foot of sorts); the tarsus ends in a claw made up of either two or three points, depending on the family to which the spider belongs. Although all arthropods use muscles attached to the inside of the exoskeleton to flex their limbs, spiders and a few other groups still use hydraulic pressure to extend them, a system inherited from their pre-arthropod ancestors. The only extensor muscles in spider legs are located in the three hip joints (bordering the coxa and the trochanter). As a result, a spider with a punctured cephalothorax cannot extend its legs, and the legs of dead spiders curl up. Spiders can generate pressures up to eight times their resting level to extend their legs, and jumping spiders can jump up to 50 times their own length by suddenly increasing the blood pressure in the third or fourth pair of legs. Although larger spiders use hydraulics to straighten their legs, unlike smaller jumping spiders they depend on their flexor muscles to generate the propulsive force for their jumps.

Most spiders that hunt actively, rather than relying on webs, have dense tufts of fine hairs between the paired claws at the tips of their legs. These tufts, known as scopulae, consist of bristles whose ends are split into as many as 1,000 branches, and enable spiders with scopulae to walk up vertical glass and upside down on ceilings. It appears that scopulae get their grip from contact with extremely thin layers of water on surfaces. Spiders, like most other arachnids, keep at least four legs on the surface while walking or running.

Silk Production

An orb weaver producing silk from its spinnerets.

The abdomen has no appendages except those that have been modified to form one to four (usually three) pairs of short, movable spinnerets, which emit silk. Each spinneret has many spigots, each of which is connected to one silk gland. There are at least six types of silk gland, each producing a different type of silk.

Silk is mainly composed of a protein very similar to that used in insect silk. It is initially a liquid, and hardens not by exposure to air but as a result of being drawn out, which changes the internal structure of the protein. It is similar in tensile strength to nylon and biological materials such as chitin, collagen and cellulose, but is much more elastic. In other words, it can stretch much further before breaking or losing shape.

Some spiders have a cribellum, a modified spinneret with up to 40,000 spigots, each of which produces a single very fine fiber. The fibers are pulled out by the calamistrum, a comb-like set of bristles on the jointed tip of the cribellum, and combined into a composite woolly thread that is very effective in snagging the bristles of insects. The earliest spiders had cribella, which produced the first silk capable of capturing insects, before spiders developed silk coated with sticky droplets. However, most modern groups of spiders have lost the cribellum.

Even species that do not build webs to catch prey use silk in several ways: as wrappers for sperm and for fertilized eggs; as a "safety rope"; for nest-building; and as "parachutes" by the young of some species.

Reproduction and Life Cycle

Spiders reproduce sexually and fertilization is internal but indirect, in other words the sperm is not inserted into the female's body by the male's genitals but by an intermediate stage. Unlike many land-living arthropods, male spiders do not produce ready-made spermatophores (packages of sperm), but spin small sperm webs on to which they ejaculate and then transfer the sperm to special syringe-like structures, palpal bulbs or palpal organs, borne on the tips of the pedipalps of mature males. When a male detects signs of a female nearby he checks whether she is of the same species and whether she is ready to mate; for example in species that produce webs or "safety ropes", the male can identify the species and sex of these objects by "smell".

Spiders generally use elaborate courtship rituals to prevent the large females from eating the small males before fertilization, except where the male is so much smaller that he is not worth eating. In web-weaving species, precise patterns of vibrations in the web are a major part of the rituals, while patterns of touches on the female's body are important in many spiders that hunt actively, and may "hypnotize" the female. Gestures and dances by the male are important for jumping spiders, which have excellent eyesight. If courtship is successful, the male injects his sperm from the palpal bulbs into the female via one or two openings on the underside of her abdomen.

Spider Fertilization systems

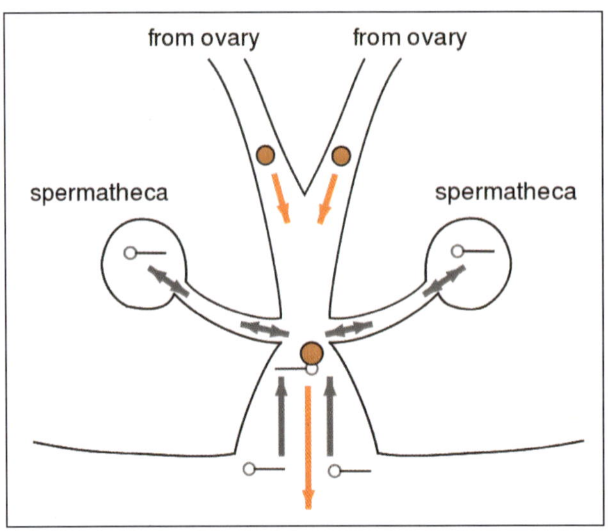

Haplogyne or non-entelegyne.

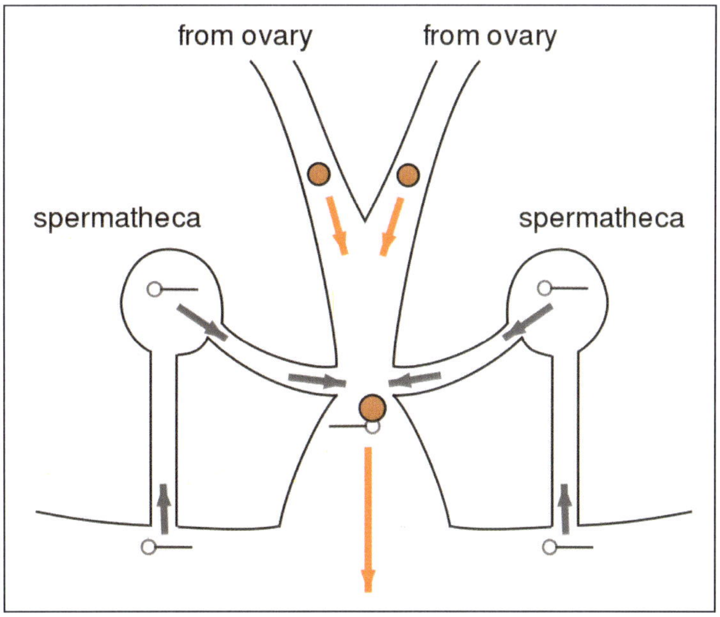

Entelegyne
Schematic diagrams showing sperm entering and being stored in the spermathecae; eggs leaving the ovaries and being fertilized; and finally a fertilized egg leaving the female's body.

Female spiders' reproductive tracts are arranged in one of two ways. The ancestral arrangement ("haplogyne" or "non-entelegyne") consists of a single genital opening, leading to two seminal receptacles (spermathecae) in which females store sperm. In the more advanced arrangement ("entelegyne"), there are two further openings leading directly to the spermathecae, creating a "flow through" system rather than a "first-in first-out" one. Eggs are as a general rule only fertilized during oviposition when the stored sperm is released from its chamber, rather than in the ovarian cavity. A few exceptions exist, such as *Parasteatoda tepidariorum*. In these species the female appears to be able to activate the dormant sperm before oviposition, allowing them to migrate to the ovarian cavity where fertilization occurs. The only known example of direct fertilization between male and female is an Israeli spider, *Harpactea sadistica*, which has evolved traumatic insemination. In this species the male will penetrate its pedipalps through the female's body wall and inject his sperm directly into her ovaries, where the embryos inside the fertilized eggs will start to develop before being laid.

Males of the genus *Tidarren* amputate one of their palps before maturation and enter adult life with one palp only. The palps are 20% of male's body mass in this species, and detaching one of the two improves mobility. In the Yemeni species *Tidarren argo*, the remaining palp is then torn off by the female. The separated palp remains attached to the female's epigynum for about four hours and apparently continues to function independently. In the meantime, the female feeds on the palpless male. In over 60% of cases, the female of the Australian redback spider kills and eats the male after it inserts its second palp into the female's genital opening; in fact, the males co-operate by trying to impale themselves on the females' fangs. Observation shows that most male redbacks never get an opportunity to mate, and the "lucky" ones increase the likely number of offspring by ensuring that the females are well-fed. However, males of most species survive a few matings, limited mainly by their short life spans. Some even live for a while in their mates' webs.

The tiny male of the Golden orb weaver (*Nephila clavipes*) (near the top of the leaf) is protected from the female by producing the right vibrations in the web, and may be too small to be worth eating.

Orange spider egg sac hanging from ceiling.

Gasteracantha mammosa spiderlings next to their eggs capsule.

Females lay up to 3,000 eggs in one or more silk egg sacs, which maintain a fairly constant humidity level. In some species, the females die afterwards, but females of other species protect the sacs by attaching them to their webs, hiding them in nests, carrying them in the chelicerae or attaching them to the spinnerets and dragging them along.

Baby spiders pass all their larval stages inside the egg and hatch as spiderlings, very small and sexually immature but similar in shape to adults. Some spiders care for their young, for example a wolf spider's brood cling to rough bristles on the mother's back, and females of some species respond to the "begging" behavior of their young by giving them their prey, provided it is no longer struggling, or even regurgitate food.

Like other arthropods, spiders have to molt to grow as their cuticle ("skin") cannot stretch. In some species males mate with newly molted females, which are too weak to be dangerous to the males. Most spiders live for only one to two years, although some tarantulas can live in captivity for over 20 years, and an Australian female trapdoor spider was documented to have lived in the wild for 43 years, dying of a parasitic wasp attack.

Wolf spider carrying its young on its abdomen.

Size

Spiders occur in a large range of sizes. The smallest, *Patu digua* from Colombia, are less than 0.37 mm (0.015 in) in body length. The largest and heaviest spiders occur among tarantulas, which can have body lengths up to 90 mm (3.5 in) and leg spans up to 250 mm (9.8 in).

Goliath birdeater (*Theraphosa blondi*), the largest spider.

Coloration

Only three classes of pigment (ommochromes, bilins and guanine) have been identified in spiders, although other pigments have been detected but not yet characterized. Melanins, carotenoids and pterins, very common in other animals, are apparently absent. In some species, the exocuticle of the legs and prosoma is modified by a tanning process, resulting in brown coloration. Bilins are found, for example, in *Micrommata virescens*, resulting in its green color. Guanine is responsible for the white markings of the European garden spider *Araneus diadematus*. It is in many species accumulated in specialized cells called guanocytes. In genera such as *Tetragnatha*, *Leucauge*, *Argyrodes* or *Theridiosoma*, guanine creates their silvery appearance. While guanine is originally an end-product of protein metabolism, its excretion can be blocked in spiders, leading to an increase in its storage. Structural colors occur in some species, which are the result of the diffraction, scattering or interference of light, for example by modified setae or scales. The white prosoma of *Argiope* results from hairs reflecting the light, *Lycosa* and *Josa* both have areas of modified cuticle that act as light reflectors.

Ecology and Behavior

Non-predatory Feeding

A jumping spider seen in Chennai.

Although spiders are generally regarded as predatory, the jumping spider *Bagheera kiplingi* gets over 90% of its food from fairly solid plant material produced by acacias as part of a mutually beneficial relationship with a species of ant.

Juveniles of some spiders in the families Anyphaenidae, Corinnidae, Clubionidae, Thomisidae and Salticidae feed on plant nectar. Laboratory studies show that they do so deliberately and over extended periods, and periodically clean themselves while feeding. These spiders also prefer sugar solutions to plain water, which indicates that they are seeking nutrients. Since many spiders are nocturnal, the extent of nectar consumption by spiders may have been underestimated. Nectar contains amino acids, lipids, vitamins and minerals in addition to sugars, and studies have shown that other spider species live longer when nectar is available. Feeding on nectar avoids the risks of struggles with prey, and the costs of producing venom and digestive enzymes.

Various species are known to feed on dead arthropods (scavenging), web silk, and their own shed exoskeletons. Pollen caught in webs may also be eaten, and studies have shown that young spiders have a better chance of survival if they have the opportunity to eat pollen. In captivity, several spider species are also known to feed on bananas, marmalade, milk, egg yolk and sausages.

Capturing Prey

The *Phonognatha graeffei* or leaf-curling spider's web serves both as a trap and as a way of making its home in a leaf.

The best-known method of prey capture is by means of sticky webs. Varying placement of webs allows different species of spider to trap different insects in the same area, for example flat horizontal webs trap insects that fly up from vegetation underneath while flat vertical webs trap insects in horizontal flight. Web-building spiders have poor vision, but are extremely sensitive to vibrations.

Females of the water spider *Argyroneta aquatica* build underwater "diving bell" webs that they fill with air and use for digesting prey, molting, mating and raising offspring. They live almost entirely within the bells, darting out to catch prey animals that touch the bell or the threads that anchor it. A few spiders use the surfaces of lakes and ponds as "webs", detecting trapped insects by the vibrations that these cause while struggling.

Net-casting spiders weave only small webs, but then manipulate them to trap prey. Those of the genus *Hyptiotes* and the family Theridiosomatidae stretch their webs and then release them when prey strike them, but do not actively move their webs. Those of the family Deinopidae weave even smaller webs, hold them outstretched between their first two pairs of legs, and lunge and push the webs as much as twice their own body length to trap prey, and this move may increase the webs' area by a factor of up to ten. Experiments have shown that *Deinopis spinosus* has two different techniques for trapping prey: backwards strikes to catch flying insects, whose vibrations it detects; and forward strikes to catch ground-walking prey that it sees. These two techniques have also been observed in other deinopids. Walking insects form most of the prey of most deinopids, but one population of *Deinopis subrufa* appears to live mainly on tipulid flies that they catch with the backwards strike.

Mature female bolas spiders of the genus *Mastophora* build "webs" that consist of only a single "trapeze line", which they patrol. They also construct a bolas made of a single thread, tipped with a large ball of very wet sticky silk. They emit chemicals that resemble the pheromones of moths, and then swing the bolas at the moths. Although they miss on about 50% of strikes, they catch about the same weight of insects per night as web-weaving spiders of similar size. The spiders eat the bolas if they have not made a kill in about 30 minutes, rest for a while, and then make new bolas. Juveniles and adult males are much smaller and do not make bolas. Instead they release different pheromones that attract moth flies, and catch them with their front pairs of legs.

A trapdoor spider in the genus *Cyclocosmia*, an ambush predator.

The primitive Liphistiidae, the "trapdoor spiders" of the family Ctenizidae and many tarantulas are ambush predators that lurk in burrows, often closed by trapdoors and often surrounded by networks of silk threads that alert these spiders to the presence of prey. Other ambush predators do without such aids, including many crab spiders, and a few species that prey on bees, which see ultraviolet, can adjust their ultraviolet reflectance to match the flowers in which they are lurking.

Wolf spiders, jumping spiders, fishing spiders and some crab spiders capture prey by chasing it, and rely mainly on vision to locate prey.

Portia uses both webs and cunning, versatile tactics to overcome prey.

Some jumping spiders of the genus *Portia* hunt other spiders in ways that seem intelligent, outflanking their victims or luring them from their webs. Laboratory studies show that *Portia*'s instinctive tactics are only starting points for a trial-and-error approach from which these spiders learn very quickly how to overcome new prey species. However, they seem to be relatively slow "thinkers", which is not surprising, as their brains are vastly smaller than those of mammalian predators.

An ant-mimicking jumping spider.

Ant-mimicking spiders face several challenges: they generally develop slimmer abdomens and false "waists" in the cephalothorax to mimic the three distinct regions (tagmata) of an ant's body; they wave the first pair of legs in front of their heads to mimic antennae, which spiders lack, and to conceal the fact that they have eight legs rather than six; they develop large color patches round one pair of eyes to disguise the fact that they generally have eight simple eyes, while ants have two compound eyes; they cover their bodies with reflective hairs to resemble the shiny bodies of ants. In some spider species, males and females mimic different ant species, as female spiders are usually much larger than males. Ant-mimicking spiders also modify their behavior to resemble that of the target species of ant; for example, many adopt a zig-zag pattern of movement, ant-mimicking jumping spiders avoid jumping, and spiders of the genus *Synemosyna* walk on the outer edges of leaves in the same way as *Pseudomyrmex*. Ant-mimicry in many spiders and other arthropods may be for protection from predators that hunt by sight, including birds, lizards and spiders. However, several ant-mimicking spiders prey either on ants or on the ants' "livestock", such as aphids. When at rest, the ant-mimicking crab spider *Amyciaea* does not closely resemble *Oecophylla*, but while hunting it imitates the behavior of a dying ant to attract worker ants. After a kill, some ant-mimicking spiders hold their victims between themselves and large groups of ants to avoid being attacked.

Defense

Threat display by a Sydney funnel-web spider (*Atrax robustus*).

There is strong evidence that spiders' coloration is camouflage that helps them to evade their major predators, birds and parasitic wasps, both of which have good color vision. Many spider species are colored so as to merge with their most common backgrounds, and some have disruptive coloration, stripes and blotches that break up their outlines. In a few species, such as the Hawaiian happy-face spider, *Theridion grallator*, several coloration schemes are present in a ratio that appears to remain constant, and this may make it more difficult for predators to recognize the species. Most spiders are insufficiently dangerous or unpleasant-tasting for warning coloration to offer much benefit. However, a few species with powerful venoms, large jaws or irritant hairs have patches of warning colors, and some actively display these colors when threatened.

Many of the family *Theraphosidae*, which includes tarantulas and baboon spiders, have urticating hairs on their abdomens and use their legs to flick them at attackers. These hairs are fine setae (bristles) with fragile bases and a row of barbs on the tip. The barbs cause intense irritation but there is no evidence that they carry any kind of venom. A few defend themselves against wasps by including networks of very robust threads in their webs, giving the spider time to flee while the wasps are struggling with the obstacles. The golden wheeling spider, *Carparachne aureoflava*, of the Namibian desert escapes parasitic wasps by flipping onto its side and cartwheeling down sand dunes.

Socialization

A few spider species that build webs live together in large colonies and show social behavior, although not as complex as in social insects. *Anelosimus eximius* (in the family *Theridiidae*) can form colonies of up to 50,000 individuals. The genus *Anelosimus* has a strong tendency towards sociality: all known American species are social, and species in Madagascar are at least somewhat social. Members of other species in the same family but several different genera have independently developed social behavior. For example, although *Theridion nigroannulatum* belongs to a genus with no other social species, *T. nigroannulatum* build colonies that may contain several thousand individuals that co-operate in prey capture and share food. Other communal spiders include several *Philoponella* species (family Uloboridae), *Agelena consociata* (family Agelenidae) and *Mallos gregalis* (family Dictynidae). Social predatory spiders need to defend their prey against kleptoparasites ("thieves"), and larger colonies are more successful in this. The herbivorous spider *Bagheera kiplingi* lives in small colonies which help to protect eggs and spiderlings. Even widow spiders (genus *Latrodectus*), which are notoriously cannibalistic, have formed small colonies in captivity, sharing webs and feeding together.

Web Types

The large orb web of *Araneus diadematus* (European garden spider).

There is no consistent relationship between the classification of spiders and the types of web they build: species in the same genus may build very similar or significantly different webs. Nor is there much correspondence between spiders' classification and the chemical composition of their silks. Convergent evolution in web construction, in other words use of similar techniques by remotely related species, is rampant. Orb web designs and the spinning behaviors that produce them are the best understood. The basic radial-then-spiral sequence visible in orb webs and the sense of direction required to build them may have been inherited from the common ancestors of most spider groups. However, the majority of spiders build non-orb webs. It used to be thought that the sticky orb web was an evolutionary innovation resulting in the diversification of the Orbiculariae. Now, however, it appears that non-orb spiders are a sub-group that evolved from orb-web spiders, and non-orb spiders have over 40% more species and are four times as abundant as orb-web spiders. Their greater success may be because sphecid wasps, which are often the dominant predators of spiders, much prefer to attack spiders that have flat webs.

Orb

About half the potential prey that hit orb webs escape. A web has to perform three functions: intercepting the prey (intersection), absorbing its momentum without breaking (stopping), and trapping the prey by entangling it or sticking to it (retention). No single design is best for all prey. For example: wider spacing of lines will increase the web's area and hence its ability to intercept prey, but reduce its stopping power and retention; closer spacing, larger sticky droplets and thicker lines would improve retention, but would make it easier for potential prey to see and avoid the web, at least during the day. However, there are no consistent differences between orb webs built for use during the day and those built for use at night. In fact, there is no simple relationship between orb web design features and the prey they capture, as each orb-weaving species takes a wide range of prey.

Nephila clavata, a golden orb weaver.

The hubs of orb webs, where the spiders lurk, are usually above the center, as the spiders can move downwards faster than upwards. If there is an obvious direction in which the spider can retreat to avoid its own predators, the hub is usually offset towards that direction.

Horizontal orb webs are fairly common, despite being less effective at intercepting and retaining prey and more vulnerable to damage by rain and falling debris. Various researchers have suggested that horizontal webs offer compensating advantages, such as reduced vulnerability to wind damage; reduced visibility to prey flying upwards, because of the back-lighting from the sky; enabling oscillations to catch insects in slow horizontal flight. However, there is no single explanation for the common use of horizontal orb webs.

Spiders often attach highly visible silk bands, called decorations or stabilimenta, to their webs. Field research suggests that webs with more decorative bands captured more prey per hour. However, a laboratory study showed that spiders reduce the building of these decorations if they sense the presence of predators.

There are several unusual variants of orb web, many of them convergently evolved, including: attachment of lines to the surface of water, possibly to trap insects in or on the surface; webs with twigs through their centers, possibly to hide the spiders from predators; "ladder-like" webs that appear most effective in catching moths. However, the significance of many variations is unclear.

In 1973, Skylab 3 took two orb-web spiders into space to test their web-spinning capabilities in zero gravity. At first, both produced rather sloppy webs, but they adapted quickly.

Cobweb

A funnel web.

Members of the family Theridiidae weave irregular, tangled, three-dimensional webs, popularly known as cobwebs. There seems to be an evolutionary trend towards a reduction in the amount of sticky silk used, leading to its total absence in some species. The construction of cobwebs is less stereotyped than that of orb-webs, and may take several days.

Other

The Linyphiidae generally make horizontal but uneven sheets, with tangles of stopping threads above. Insects that hit the stopping threads fall onto the sheet or are shaken onto it by the spider, and are held by sticky threads on the sheet until the spider can attack from below.

Evolution

Fossil Record

Spider preserved in amber.

Although the fossil record of spiders is considered poor, almost 1000 species have been described from fossils. Because spiders' bodies are quite soft, the vast majority of fossil spiders have been found preserved in amber. The oldest known amber that contains fossil arthropods dates from 130 million years ago in the Early Cretaceous period. In addition to preserving spiders' anatomy in very fine detail, pieces of amber show spiders mating, killing prey, producing silk and possibly caring for their young. In a few cases, amber has preserved spiders' egg sacs and webs, occasionally with prey attached; the oldest fossil web found so far is 100 million years old. Earlier spider fossils come from a few lagerstätten, places where conditions were exceptionally suited to preserving fairly soft tissues.

The oldest known exclusively terrestrial arachnid is the trigonotarbid *Palaeotarbus jerami*, from about 420 million years ago in the Silurian period, and had a triangular cephalothorax and segmented abdomen, as well as eight legs and a pair of pedipalps. *Attercopus fimbriunguis*, from 386 million years ago in the Devonian period, bears the earliest known silk-producing spigots, and was therefore hailed as a spider at the time of its discovery. However, these spigots may have been mounted on the underside of the abdomen rather than on spinnerets, which are modified appendages and whose mobility is important in the building of webs. Hence *Attercopus* and the similar Permian arachnid *Permarachne* may not have been true spiders, and probably used silk for lining nests or producing egg-cases rather than for building webs. The largest known fossil spider as of 2011 is the araneid *Nephila jurassica*, from about 165 million years ago, recorded from Daohuogo, Inner Mongolia in China. Its body length is almost 25 mm, (i.e., almost one inch).

Several Carboniferous spiders were members of the Mesothelae, a primitive group now represented only by the Liphistiidae. The mesothelid *Paleothele montceauensis*, from the Late Carboniferous over 299 million years ago, had five spinnerets. Although the Permian period 299 to 251 million years ago saw rapid diversification of flying insects, there are very few fossil spiders from this period.

The main groups of modern spiders, Mygalomorphae and Araneomorphae, first appear in the Triassic well before 200 million years ago. Some Triassic mygalomorphs appear to be members of the family Hexathelidae, whose modern members include the notorious Sydney funnel-web spider, and their spinnerets appear adapted for building funnel-shaped webs to catch jumping insects. Araneomorphae account for the great majority of modern spiders, including those that weave the familiar orb-shaped webs. The Jurassic and Cretaceous periods provide a large number of fossil spiders, including representatives of many modern families.

Family Tree

It is now agreed that spiders (Araneae) are monophyletic (i.e., members of a group of organisms that form a clade, consisting of a last common ancestor and all of its descendants). There has been debate about what their closest evolutionary relatives are, and how all of these evolved from the ancestral chelicerates, which were marine animals. The cladogram on the right is based on J. W. Shultz' analysis. Other views include proposals that: scorpions are more closely related to the extinct marine scorpion-like eurypterids than to spiders; spiders and Amblypygi are a monophyletic group. The appearance of several multi-way branchings in the tree on the right shows that there are still uncertainties about relationships between the groups involved.

Arachnids lack some features of other chelicerates, including backward-pointing mouths and gnathobases ("jaw bases") at the bases of their legs; both of these features are part of the ancestral arthropod feeding system. Instead, they have mouths that point forwards and downwards, and all have some means of breathing air. Spiders (Araneae) are distinguished from other arachnid groups by several characteristics, including spinnerets and, in males, pedipalps that are specially adapted for sperm transfer.

Taxonomy

Spiders are divided into two suborders, Mesothelae and Opisthothelae, of which the latter contains two infraorders, Mygalomorphae and Araneomorphae. Over 48,000 living species of spiders (order Araneae) have been identified and as of 2019 grouped into 120 families and about 4,100 genera by arachnologists.

<table>
<tr><td></td><td colspan="3">Spider diversity
(numbers are approximate)</td><td colspan="4">Features</td></tr>
<tr><td>Suborder/Infraorder</td><td>Families</td><td>Genera</td><td>Species</td><td>Segmented plates on top of abdomen</td><td>Ganglia in abdomen</td><td>Spinnerets</td><td>Striking direction of fangs</td></tr>
<tr><td>Mesothelae</td><td>1</td><td>8</td><td>116</td><td>Yes</td><td>Yes</td><td>Four pairs, in some species one pair fused, under middle of abdomen</td><td rowspan="2">Downwards and forwards</td></tr>
<tr><td>Opisthothelae: Mygalomorphae</td><td>20</td><td>350</td><td>2,900</td><td rowspan="2">Only in some fossils</td><td rowspan="2">No</td><td rowspan="2">One, two or three pairs under rear of abdomen</td></tr>
<tr><td>Opisthothelae: Araneomorphae</td><td>96</td><td>3,700</td><td>44,000</td><td>From sides to center, like pincers</td></tr>
</table>

Mesothelae

Ryuthela sasakii, a member of the Liphistiidae

The only living members of the primitive Mesothelae are the family Liphistiidae, found only in Southeast Asia, China, and Japan. Most of the Liphistiidae construct silk-lined burrows with thin trapdoors, although some species of the genus *Liphistius* build camouflaged silk tubes with a second trapdoor as an emergency exit. Members of the genus *Liphistius* run silk "tripwires" outwards from their tunnels to help them detect approaching prey, while those of genus *Heptathela* do not and instead rely on their built-in vibration sensors. Spiders of the genus *Heptathela* have no venom glands although they do have venom gland outlets on the fang tip.

The extinct families Arthrolycosidae, found in Carboniferous and Permian rocks, and Arthromygalidae, so far found only in Carboniferous rocks, have been classified as members of the Mesothelae.

Mygalomorphae

A Mexican red-kneed tarantula *Brachypelma smithi.*

The Mygalomorphae, which first appeared in the Triassic period, are generally heavily built and hairy, with large, robust chelicerae and fangs. Well-known examples include tarantulas, ctenizid trapdoor spiders and the Australasian funnel-web spiders. Most spend the majority of their time in burrows, and some run silk tripwires out from these, but a few build webs to capture prey. However, mygalomorphs cannot produce the pirifom silk that the Araneomorphae use as instant adhesive to glue silk to surfaces or to other strands of silk, and this makes web construction more difficult for mygalomorphs. Since mygalomorphs rarely "balloon" by using air currents for transport, their populations often form clumps. In addition to arthropods, mygalomorphs are capable of preying on frogs, small mammals, lizards, and snails.

Araneomorphae

Leucauge venusta, an orb-web spider.

In addition to accounting for over 90% of spider species, the Araneomorphae, also known as the "true spiders", include orb-web spiders, the cursorial wolf spiders, and jumping spiders, as well as the only known herbivorous spider, *Bagheera kiplingi.* They are distinguished by having fangs that oppose each other and cross in a pinching action, in contrast to the Mygalomorphae, which have fangs that are nearly parallel in alignment.

Human Interaction

Bites

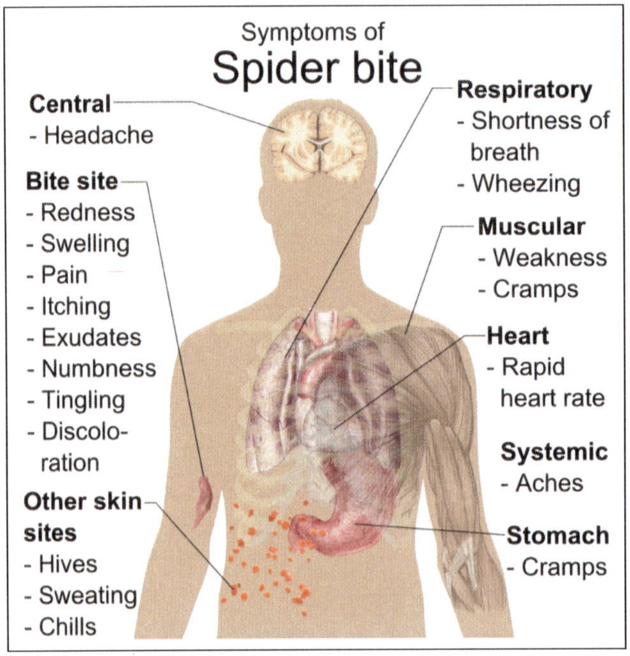

All symptoms associated with toxic spider bites.

Although spiders are widely feared, only a few species are dangerous to people. Spiders will only bite humans in self-defense, and few produce worse effects than a mosquito bite or bee-sting. Most of those with medically serious bites, such as recluse spiders (genus *Loxosceles*) and widow spiders (genus *Latrodectus*), would rather flee and bite only when trapped, although this can easily arise by accident. The defensive tactics of Australian funnel-web spiders (family Atracidae) include fang display. Their venom, although they rarely inject much, has resulted in 13 attributed human deaths over 50 years. They have been deemed to be the world's most dangerous spiders on clinical and venom toxicity grounds, though this claim has also been attributed to the Brazilian wandering spider (genus *Phoneutria*).

There were about 100 reliably reported deaths from spider bites in the 20th century, compared to about 1,500 from jellyfish stings. Many alleged cases of spider bites may represent incorrect diagnoses, which would make it more difficult to check the effectiveness of treatments for genuine bites. A review published in 2016 agreed with this conclusion, showing that 78% of 134 published medical case studies of supposed spider bites did not meet the necessary criteria for a spider bite to be verified. In the case of the two genera with the highest reported number of bites, *Loxosceles* and *Latrodectus*, spider bites were not verified in over 90% of the reports. Even when verification had occurred, details of the treatment and its effects were often lacking.

Chemical Benefits

Spider venoms may be a less polluting alternative to conventional pesticides, as they are deadly to insects but the great majority are harmless to vertebrates. Australian funnel web spiders are a promising source, as most of the world's insect pests have had no opportunity to develop any immunity to their venom, and funnel web spiders thrive in captivity and are easy to "milk". It may be possible to target specific pests by engineering genes for the production of spider toxins into viruses that infect species such as cotton bollworms.

The Ch'ol Maya use a beverage created from the tarantula species *Brachypelma vagans* for the treatment of a condition they term 'tarantula wind', the symptoms of which include chest pain, asthma and coughing.

Possible medical uses for spider venoms are being investigated, for the treatment of cardiac arrhythmia, Alzheimer's disease, strokes, and erectile dysfunction. The peptide GsMtx-4, found in the venom of Brachypelma vagans, is being researched to determine whether or not it could effectively be used for the treatment of cardiac arrhythmia, muscular dystrophy or glioma. Because spider silk is both light and very strong, attempts are being made to produce it in goats' milk and in the leaves of plants, by means of genetic engineering.

Spiders can also be used as food. Cooked tarantula spiders are considered a delicacy in Cambodia, and by the Piaroa Indians of southern Venezuela – provided the highly irritant hairs, the spiders' main defense system, are removed first.

Cooked tarantula spiders are considered a delicacy in Cambodia.

Arachnophobia

Arachnophobia is a specific phobia—it is the abnormal fear of spiders or anything reminiscent of spiders, such as webs or spider-like shapes. It is one of the most common specific phobias, and some statistics show that 50% of women and 10% of men show symptoms. It may be an exaggerated form of an instinctive response that helped early humans to survive, or a cultural phenomenon that is most common in predominantly European societies.

Spiders in Culture

Spiders have been the focus of stories and mythologies of various cultures for centuries. Uttu, the ancient Sumerian goddess of weaving, was envisioned as a spider spinning her web. According to her main myth, she resisted her father Enki's sexual advances by ensconcing herself in her web, but let him in after he promised her fresh produce as a marriage gift, thereby allowing him to intoxicate her with beer and rape her. Enki's wife Ninhursag heard Uttu's screams and rescued her, removing Enki's semen from her vagina and planting it in the ground to produce eight previously nonexistent plants. In a story told by the Roman poet Ovid in his *Metamorphoses*, Arachne was a Lydian girl who challenged the goddess Athena to a weaving contest. Arachne won, but Athena destroyed her tapestry out of jealousy, causing Arachne to hang herself. In an act of mercy, Athena brought Arachne back to life as the first spider. Stories about the trickster-spider Anansi are prominent in the folktales of West Africa and the Caribbean.

In some cultures, spiders have symbolized patience due to their hunting technique of setting webs and waiting for prey, as well as mischief and malice due to their venomous bites. The Italian tarantella is a dance to rid the young woman of the lustful effects of a spider bite. Web-spinning also caused the association of the spider with creation myths, as they seem to have the ability to produce their own worlds. Dreamcatchers are depictions of spiderwebs. The Moche people of ancient Peru worshipped nature. They placed emphasis on animals and often depicted spiders in their art.

This Moche ceramic depicts a spider, and dates from around 300 CE.

Scorpiology

Scorpiology is the study of scorpions, the animals in the class Arachnida, order Scorpiones. It is a subfield of arachnology, a subdiscipline of the field of zoology.

Scorpion

Scorpions are predatory arachnids of the order Scorpiones. They have eight legs and are easily recognized by the pair of grasping pedipalps and the narrow, segmented tail, often carried in a characteristic forward curve over the back, ending with a venomous stinger. Scorpions range in size from 9 mm / 0.3 in. (*Typhlochactas mitchelli*) to 23 cm / 9 in. (*Heterometrus swammerdami*).

The evolutionary history of scorpions goes back to the Silurian period 430 million years ago. They have adapted to a wide range of environmental conditions, and they can now be found on all continents except Antarctica. Scorpions number about 1,750 described species, with 13 extant (living) families recognised to date. The taxonomy has undergone changes and is likely to change further, as genetic studies are bringing forth new information.

All scorpions have a venomous sting, but the vast majority of the species do not represent a serious threat to humans, and in most cases, healthy adults do not need any medical treatment after being stung. Only about 25 species are known to have venom capable of killing a human. In some parts of the world with highly venomous species, human fatalities regularly occur, primarily in areas with limited access to medical treatment.

Geographical Distribution

Scorpions are found on all major land masses except Antarctica and New Zealand. Scorpions did not occur naturally in Great Britain, Ireland, Japan, South Korea, and some of the islands in Oceania, but now have been accidentally introduced in some of these places by human trade and commerce. The greatest diversity of scorpions in the Northern Hemisphere is to be found in regions between the latitudes 23 and 38 °N. Above these latitudes, the diversity decreases with the northernmost natural occurrence of scorpions being the northern scorpion *Paruroctonus boreus* at Medicine Hat, Alberta, Canada 50 °N. Five colonies of scorpions (*Euscorpius flavicaudis*) have established themselves in Sheerness on the Isle of Sheppey in the United Kingdom. This small population has been resident since the 1860s, having probably arrived with imported fruit from Africa. This scorpion species is small and completely harmless to humans. At just over 51 °N, this marks the northernmost limit where scorpions live in the wild.

Today, scorpions are found in virtually every terrestrial habitat including: high-elevation mountains, caves, and intertidal zones, with the exception of boreal ecosystems such as: the tundra, high-altitude taiga, and the permanently snow-clad tops of some mountains. As regards microhabitats, scorpions may be ground-dwelling, tree-living, rock-loving or sand-loving. Some species, such as *Vaejovis janssi*, are versatile and are found in every type of habitat in Baja California, while others occupy specialized niches such as *Euscorpius carpathicus*, which is endemic to the littoral zone of rivers in Romania.

Classification

Thirteen families and about 1,750 described species and subspecies of scorpions are known. In addition, 111 described taxa of scorpions are extinct.

This classification is based on that of Soleglad and Fet, which replaced the older, unpublished classification of Stockwell. Additional taxonomic changes are from papers by Soleglad et al.

Systematics

This classification covers extant taxa to the rank of family:

Order Scorpiones

- Infraorder Orthosterni Pocock, 1911
    - Parvorder Pseudochactida Soleglad et Fet, 2003
        - Superfamily Pseudochactoidea Gromov, 1998
            - Family Pseudochactidae Gromov, 1998
    - Parvorder Buthida Soleglad et Fet, 2003
        - Superfamily Buthoidea C. L. Koch, 1837
            - Family Buthidae C. L. Koch, 1837 (thick-tailed scorpions)
            - Family Microcharmidae Lourenço, 1996

- Parvorder Chaerilida Soleglad et Fet, 2003
  - Superfamily Chaeriloidea Pocock, 1893
    - Family Chaerilidae Pocock, 1893
- Parvorder Iurida Soleglad et Fet, 2003
  - Superfamily Chactoidea Pocock, 1893
    - Family Chactidae Pocock, 1893
    - Family Euscorpiidae Laurie, 1896
    - Family Superstitioniidae Stahnke, 1940
    - Family Vaejovidae Thorell, 1876
  - Superfamily Iuroidea Thorell, 1876
    - Family Caraboctonidae Kraepelin, 1905 (hairy scorpions)
    - Family Iuridae Thorell, 1876
  - Superfamily Scorpionoidea Latreille, 1802
    - Family Bothriuridae Simon, 1880
    - Family Hemiscorpiidae Pocock, 1893 (= Ischnuridae, =Liochelidae) (rock scorpions, creeping scorpions, or tree scorpions)
    - Family Scorpionidae Latreille, 1802 (burrowing scorpions or pale-legged scorpions)

Fossil Record

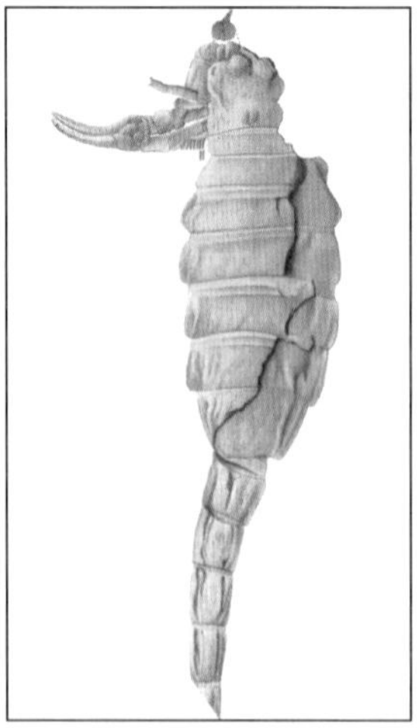

Proscorpius osborni, fossil scorpion initially thought to be a eurypterid.

Scorpion remains have been found in many fossil records, including marine Silurian and estuarine Devonian deposits, coal deposits from the Carboniferous Period and in amber. The oldest known scorpions lived around 430 million years ago in the Silurian period. Though once believed to have lived on the bottom of shallow tropical seas, early scorpions are now believed to have been terrestrial and to have washed into marine settings together with plant matter. These first scorpions were believed to have had gills instead of the present forms' book lungs, though this has subsequently been refuted. The oldest Gondwanan scorpions (*Gondwanascorpio*) comprise the earliest known terrestrial animals from Gondwana. Currently, 111 fossil species of scorpion are known. Unusually for arachnids, there are more species of Palaeozoic scorpion than Mesozoic or Cenozoic ones.

Ancestral scorpions had compound eyes, but as they adapted to a nocturnal lifestyle, they became simplified.

The eurypterids, commonly called "sea scorpions", were aquatic creatures that lived during the Palaeozoic era that share several physical traits with scorpions and may be closely related to them. Various species of Eurypterida could grow to be anywhere from 10 centimetres (3.9 in) to 2.5 metres (8.2 ft) in length. However, they exhibit anatomical differences marking them off as a group distinct from their Carboniferous and Recent relatives. Despite this, they are commonly referred to as "sea scorpions". Their legs are thought to have been short, thick, tapering and to have ended in a single strong claw. It appears that they were well-adapted for maintaining a secure hold upon rocks or seaweed against the wash of waves, like the legs of a shore crab. Cladistic analyses have supported the idea that the eurypterids are a distinct group from the scorpions.

Morphology

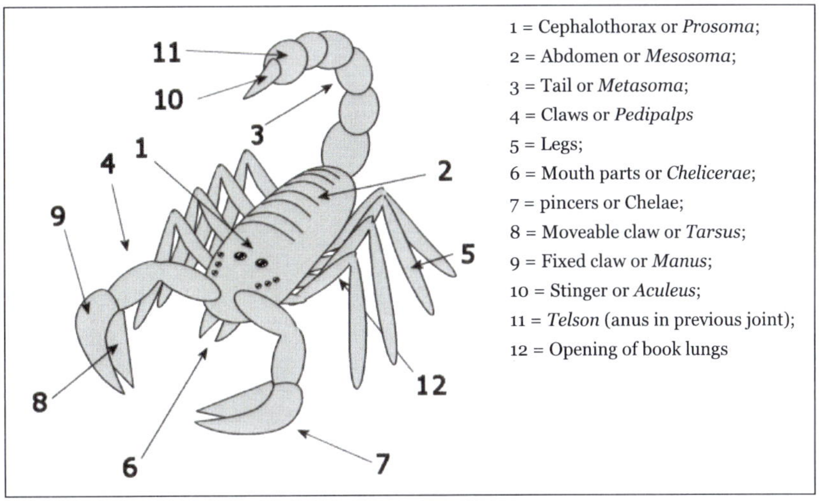

Scorpion anatomy.

Stinger of an Arizona bark scorpion.

Ventral view of an unidentified scorpion species where the pectines can easily be observed with a comb like structure in an inverted V shape.

The body of a scorpion is divided into two parts (tagmata): the head (cephalothorax) and the abdomen (opisthosoma), which is subdivided into a broad anterior (mesosoma), or preabdomen, and a narrow tail-like posterior (metasoma), or postabdomen.

Cephalothorax

The cephalothorax, also called the *prosoma*, comprises the carapace, eyes, chelicerae (mouth parts), pedipalps (the pedipalps of scorpions have chelae, commonly called claws or pincers) and four pairs of walking legs. The scorpion's exoskeleton is thick and durable, providing good protection from predators. Scorpions have two eyes on the top of the cephalothorax, and usually two to five pairs of eyes along the front corners of the cephalothorax. While unable to form sharp images, their central eyes are amongst the most light sensitive in the animal kingdom, especially in dim light, and makes it possible for nocturnal species to use star light to navigate at night. Some species also have light receptors in their tail. The position of the eyes on the cephalothorax depends in part on the hardness or softness of the soil upon which they spend their lives.

The pedipalp is a segmented, chelate (clawed) appendage used for prey immobilization, defense and sensory purposes. The segments of the pedipalp (from closest to the body outwards) are coxa, trochanter, femur (humerus), patella, tibia (including the fixed claw and the manus) and tarsus (moveable claw). A scorpion has darkened or granular raised linear ridges, called "keels" or *carinae* on the pedipalp segments and on other parts of the body, which are useful taxonomically.

Mesosoma

The mesosoma is the broad part of the opisthosoma. Sometimes it is loosely called the *abdomen*. It consists of the anterior seven somites (segments) of the opisthosoma, each covered dorsally by a sclerotosed plate, its tergite. Ventrally somites 3 to 7 are armoured with matching plates called sternites.

Ventrally somites 1 and 2 are more complex; the first abdominal sternite is modified into a pair of genital opercula covering the gonopore. Sternite 2 forms the basal plate bearing the pectines. Morphologically the pectines are a pair of limbs that function as sensory organs.

The next four somites, 3 to 6, all bear pairs of spiracles. They serve as openings for the scorpion's respiratory organs, known as book lungs. The spiracle openings may be slits, circular, elliptical or oval according to the species of scorpion.

The 7th and last somite do not bear appendages or any other significant external structures.

Metasoma

The metasoma is commonly known as the scorpion's "tail", though this is in some ways misleading because unlike most so-called tails it is not an appendage or limb. It is in fact part of the opisthosoma. It comprises five segments, of which the fifth segment bears the telson. In many species, it superficially seems as though the metasoma has four segments only, because their first (anterior) metasomal segment gives the impression of being the posterior segment of the mesosoma. The fifth segment of the metasoma is the caudal segment of the opisthosoma, and accordingly bears the anus. The scorpion's telson is the part commonly called the stinger; it is attached to the end of the fifth segment just dorsad from the anus, but as the distal end of the tail at rest normally is carried upside down with the sting pointing forward, the anus usually is above the base of the telson and facing upwards.

The telson includes the vesicle, containing a symmetrical pair of venom glands. Externally it bears the curved sting, the hypodermic aculeus or venom-injecting barb. It is equipped with various sensory hairs, as the sting cannot be guided visually. Each of the venom glands has its own duct to convey its secretion internally along the aculeus from the bulb of the gland to immediately subterminal of the point of the aculeus, where each of the paired ducts has its own venom pore.

On rare occasions, scorpions are born with two metasomata. Two-tailed scorpions are no more than examples of adventitious ontogenic abnormality. Whether there ever is a genetic component to the condition is uncertain, but such evidence as is available from offspring is negative so far as no two-tailed scorpions have been observed among the rarely-observed progeny of multiple-tailed scorpion specimens.

Fluorescence

This black scorpion appears light-blue under a black light.

Scorpions are also known to glow a vibrant blue-green when exposed to certain wavelengths of ultraviolet light such as that produced by a black light, due to the presence of fluorescent chemicals in the cuticle. One fluorescent component is now known to be beta-carboline. A hand-held UV lamp has long been a standard tool for nocturnal field surveys of these animals. Fluorescence occurs as a result of sclerotisation and increases in intensity with each successive instar. This fluorescence may have an active role in scorpion light detection.

Biology

Scorpions prefer areas where the temperatures range from 20 to 37 °C (68 to 99 °F), but may survive temperatures ranging from well below freezing to desert heat. Scorpions of the genus *Scorpiops* living in high Asian mountains, bothriurid scorpions from Patagonia and small *Euscorpius* scorpions from Central Europe can all survive winter temperatures of about −25 °C (−13 °F). In Repetek (Turkmenistan), seven species of scorpion (of which *Pectinibuthus birulai* is endemic) live in temperatures varying from −31 to 50 °C (−24 to 122 °F).

They are nocturnal and fossorial, finding shelter during the day in the relative cool of underground holes or undersides of rocks, and emerging at night to hunt and feed. Scorpions exhibit photophobic behavior, primarily to evade detection by predators such as birds, lizards, rodents such as the grasshopper mouse, opossums, and larger mammals including mongooses and the honey badger.

Scorpions are opportunistic predators of small arthropods, although the larger kinds have been known to kill small lizards and snakes. The large pincers are studded with highly sensitive tactile hair, and the moment an insect touches these, they use their chelae (pincers) to catch the prey. Depending on the toxicity of their venom and size of their claws, they will then either crush the prey or inject it with neurotoxic venom. This will kill or paralyze the prey so the scorpion can eat it. Scorpions have an unusual style of eating using chelicerae, small claw-like structures that protrude from the mouth that are unique to the Chelicerata among arthropods. The chelicerae, which are very sharp, are used to pull small amounts of food off the prey item for digestion into a *pre-oral cavity* below the chelicerae and carapace. Scorpions can ingest food only in a liquid form; they have external digestion. The digestive juices from the gut are egested onto the food and the digested food sucked in liquid form. Any solid indigestible matter (fur, exoskeleton, etc.) is trapped by setae in the pre-oral cavity and ejected by the scorpion.

Scorpions can consume huge amounts of food at one sitting. They have a very efficient food storage organ and a very low metabolic rate combined with a relatively inactive lifestyle. This enables scorpions to survive long periods when deprived of food. Some are able to survive 6 to 12 months of starvation. Scorpions excrete very little. Their waste consists mostly of insoluble nitrogenous compounds, such as xanthine, guanine and uric acid.

Reproduction

Most scorpions reproduce sexually, and most species have male and female individuals; however, some species, such as *Hottentotta hottentotta, Hottentotta caboverdensis, Liocheles australasiae, Tityus columbianus, Tityus metuendus, Tityus serrulatus, Tityus stigmurus, Tityus trivittatus* and *Tityus urugayensis*, reproduce through parthenogenesis, a process in which unfertilized eggs develop into living embryos. Parthenogenic reproduction starts following the scorpion's final molt to maturity and continues thereafter.

Pregnant scorpion.

Sexual reproduction is accomplished by the transfer of a spermatophore from the male to the female. Scorpions possess a complex courtship and mating ritual to effect this transfer. Mating starts with the male and female locating and identifying each other using a mixture of pheromones and vibrational communication. Once they have satisfied the other that they are of opposite sex and of the correct species, mating can commence.

The courtship starts with the male grasping the female's pedipalps with his own. The pair then perform a "dance" called the "*promenade à deux*". In this "dance," the male leads the female around searching for a suitable place to deposit his spermatophore. The courtship ritual can involve several other behaviors such as juddering and a cheliceral kiss, in which the male's chelicerae – pincers – grasp the female's in a smaller, more intimate version of the male's grasping the female's pedipalps, and in some cases injecting a small amount of his venom into her pedipalp or on the edge of her cephalothorax, probably as a means of pacifying the female.

When the male has identified a suitable location, he deposits the spermatophore and then guides the female over it. This allows the spermatophore to enter her genital opercula, which triggers release of the sperm, thus fertilizing the female. The mating process can take from 1 to 25+ hours, and depends on the ability of the male to find a suitable place to deposit his spermatophore. If mating continues too long, the female may lose interest, ending the process.

Once the mating is complete, the male will generally retreat quickly, for unknown reasons; sexual cannibalism is infrequent with scorpions.

Birth and Development

Compsobuthus werneri female with young.

Unlike the majority of species in the class Arachnida, which are oviparous, scorpions seem to be universally viviparous. The young are born one by one, expel the embryonic membrane, if any, and the brood is carried about on its mother's back until the young have undergone at least one molt. Before the first molt, scorplings cannot survive naturally without the mother, since they depend on her for protection and to regulate their moisture levels. Especially in species that display more advanced sociability (e.g. *Pandinus* spp.), the young/mother association can continue for an extended period of time. The size of the litter depends on the species and environmental factors, and can range from 2 to more than 100 scorplings. The average litter however, consists of around eight scorplings.

The young generally resemble their parents. Growth is accomplished by periodic shedding of the exoskeleton (ecdysis). A scorpion's developmental progress is measured in instars (how many molts it has undergone). Scorpions typically require between five and seven molts to reach maturity. Molting commences with a split in the old exoskeleton just below the edge of the carapace (at the front of the prosoma). The scorpion then emerges from this split. The pedipalps and legs are first removed from the old exoskeleton, followed eventually by the metasoma. When it emerges, the scorpion's new exoskeleton is soft, making the scorpion highly vulnerable to attack. The scorpion must constantly stretch while the new exoskeleton hardens to ensure that it can move when the hardening is complete. The process of hardening is called sclerotisation. The new exoskeleton does not fluoresce. As sclerotisation occurs, the fluorescence gradually returns.

Relationship with Humans

Although scorpions are usually not found in large numbers in densely populated urban areas, they do regularly occur in and near human habitation in all tropical parts of the world. The lack of predators, readily available shelter, and abundance of insect prey such as crickets, cockroaches, silverfishes, and earwigs have caused scorpions to thrive in industrial and residential areas. In Brazil, the number of people stung by scorpions increased from 12,000 in 2000 to 140,000 in 2018.

Sting and Venom

All known scorpion species possess venom and use it primarily to kill or paralyze their prey so that it can be eaten. The venom consists of a collection of peptides.

In general, the venom is fast-acting, allowing for effective prey capture; however, as a general rule, scorpions kill their prey with brute force if they can, as opposed to using venom, which is also used as a defense against predators. The venom is a mixture of compounds (neurotoxins, enzyme inhibitors, etc.), each not only causing a different effect, but possibly also targeting a specific animal. Each compound is made and stored in a pair of glandular sacs and is released in a quantity regulated by the scorpion itself. Of the more than one thousand known species of scorpions, only 25 have venom that is deadly to humans; most of those belong to the family Buthidae (including *Leiurus quinquestriatus*, *Hottentotta* spp., *Centruroides* spp., and *Androctonus* spp.).

Prevention

According to the United States National Institute for Occupational Safety and Health, these steps should be taken to prevent scorpion stings:

- Wear long sleeves and trousers.
- Wear leather gloves.

- Shake out clothing, bedding, bathroom towels, or shoes before using them.
- Workers with a history of severe allergic reactions to insect bites or stings should consider carrying an epinephrine autoinjector (EpiPen) and should wear a medical identification bracelet or necklace stating their allergy.

Treatment

First aid for scorpion stings is generally symptomatic. It includes strong analgesia, either systemic (opioids or paracetamol) or locally applied (such as a cold compress). Cases of very high blood pressure are treated with anxiety-relieving medications and medications which lower the blood pressure by widening the diameter of blood vessels. Scorpion envenomation with high morbidity and mortality is usually due to either excessive autonomic activity and cardiovascular toxic effects or neuromuscular toxic effects. Antivenom is the specific treatment for scorpion envenomation combined with supportive measures including vasodilators in patients with cardiovascular toxic effects and benzodiazepines when neuromuscular involvement occurs. Although rare, severe hypersensitivity reactions including anaphylaxis to scorpion antivenin (SAV) are possible.

Medical Use

The deathstalker has powerful venom.

Short-chain scorpion toxins constitute the largest group of potassium (K^+) channel-blocking peptides. An important physiological role of the KCNA3 channel, also known as K_V 1.3, is to help maintain large electrical gradients for the sustained transport of ions such as Ca^{2+} that controls T lymphocyte (T cell) proliferation. Thus K_V 1.3 blockers could be potential immunosuppressants for the treatment of autoimmune disorders (such as rheumatoid arthritis, inflammatory bowel disease, and multiple sclerosis).

The venom of *Uroplectes lineatus* is clinically important in dermatology.

Toxins being investigated include the following:

- Chlorotoxin is a 36–amino acid peptide found in the venom of the deathstalker scorpion (Leiurus quinquestriatus) that blocks small-conductance chloride channels. The fact that chlorotoxin binds preferentially to glioma cells has allowed the development of new methods, that still are under investigation, for the treatment and diagnosis of several types of cancer.

- Maurotoxin from the venom of the Tunisian Scorpio maurus.
- A number of antimicrobial peptides have also been found in the venom of Mesobuthus eupeus. Meucin-13 and meucin-18 exhibited extensive cytolytic effects on bacteria, fungi, and yeasts. Furthermore, meucin-24 and meucin-25, first identified from genetic sequences expressed in their venom gland, were shown to selectively kill Plasmodium falciparum and inhibit the development of Plasmodium berghei, both malaria parasites, but do not harm mammalian cells. These two venom-derived proteins are therefore attractive candidates for the development of anti-malarial drugs.

Production

Scorpions for use in the pharmaceutical industry are collected from the wild in Pakistan. Farmers in the Thatta District are paid about US$100 for each 40-gram scorpion, and 60-gram specimens are reported to fetch at least US$50,000. The trade is reported to be illegal but thriving.

The venom is one of the most valuable liquids by volume on earth and it costs $39 million to produce a gallon of the toxin.

Entomology

Entomology is the scientific study of insects, a branch of zoology. In the past the term "insect" was vaguer, and historically the definition of entomology included the study of terrestrial animals in other arthropod groups or other phyla, such as arachnids, myriapods, earthworms, land snails, and slugs. This wider meaning may still be encountered in informal use.

Like several of the other fields that are categorized within zoology, entomology is a taxon-based category; any form of scientific study in which there is a focus on insect-related inquiries is, by definition, entomology. Entomology therefore overlaps with a cross-section of topics as diverse as molecular genetics, behavior, biomechanics, biochemistry, systematics, physiology, developmental biology, ecology, morphology, and paleontology.

At some 1.3 million described species, insects account for more than two-thirds of all known organisms, date back some 400 million years, and have many kinds of interactions with humans and other forms of life on earth.

Coleopterology

Coleopterology is a branch of entomology, the scientific study of beetles of the order Coleoptera. Practitioners are termed coleopterists. Coleopterists have formed organizations to facilitate the study of beetles.

Beetle

Beetles are a group of insects that form the order Coleoptera, in the superorder Endopterygota. Their front pair of wings are hardened into wing-cases, elytra, distinguishing them from most other insects. The Coleoptera, with about 400,000 species, is the largest of all orders, constituting almost 40% of described insects and 25% of all known animal life-forms; new species are discovered frequently. The largest of all families, the Curculionidae (weevils) with some 83,000 member species, belongs to this order. Found in almost every habitat except the sea and the polar regions, they interact with their ecosystems in several ways: beetles often feed on plants and fungi, break down animal and plant debris, and eat other invertebrates. Some species are serious agricultural pests, such as the Colorado potato beetle, while others such as Coccinellidae (ladybirds or ladybugs) eat aphids, scale insects, thrips, and other plant-sucking insects that damage crops.

Beetles typically have a particularly hard exoskeleton including the elytra, though some such as the rove beetles have very short elytra while blister beetles have softer elytra. The general anatomy of a beetle is quite uniform and typical of insects, although there are several examples of novelty, such as adaptations in water beetles which trap air bubbles under the elytra for use while diving. Beetles are endopterygotes, which means that they undergo complete metamorphosis, with a series of conspicuous and relatively abrupt changes in body structure between hatching and becoming adult after a relatively immobile pupal stage. Some, such as stag beetles, have a marked sexual dimorphism, the males possessing enormously enlarged mandibles which they use to fight other males. Many beetles are aposematic, with bright colours and patterns warning of their toxicity, while others are harmless Batesian mimics of such insects. Many beetles, including those that live in sandy places, have effective camouflage.

Beetles are prominent in human culture, from the sacred scarabs of ancient Egypt to beetlewing art and use as pets or fighting insects for entertainment and gambling. Many beetle groups are brightly and attractively coloured making them objects of collection and decorative displays. Over 300 species are used as food, mostly as larvae; species widely consumed include mealworms and rhinoceros beetle larvae. However, the major impact of beetles on human life is as agricultural, forestry, and horticultural pests. Serious pests include the boll weevil of cotton, the Colorado potato beetle, the coconut hispine beetle, and the mountain pine beetle. Most beetles, however, do not cause economic damage and many, such as the lady beetles and dung beetles are beneficial by helping to control insect pests.

Distribution and Diversity

Beetles are by far the largest order of insects: the roughly 400,000 species make up about 40% of all insect species so far described, and about 25% of all animals. A 2015 study provided four independent estimates of the total number of beetle species, giving a mean estimate of some 1.5 million with a "surprisingly narrow range" spanning all four estimates from a minimum of 0.9 to a maximum of 2.1 million beetle species. The four estimates made use of host-specificity relationships (1.5 to 1.9 million), ratios with other taxa (0.9 to 1.2 million), plant:beetle ratios (1.2 to 1.3), and extrapolations based on body size by year of description (1.7 to 2.1 million).

Beetles are found in nearly all habitats, including freshwater and coastal habitats, wherever vegetative foliage is found, from trees and their bark to flowers, leaves, and underground near roots - even inside plants in galls, in every plant tissue, including dead or decaying ones. Tropical forest canopies have a large and diverse fauna of beetles, including Carabidae, Chrysomelidae, and Scarabaeidae.

The heaviest beetle, indeed the heaviest insect stage, is the larva of the goliath beetle, *Goliathus goliatus*, which can attain a mass of at least 115 g (4.1 oz) and a length of 11.5 cm (4.5 in). Adult male goliath beetles are the heaviest beetle in its adult stage, weighing 70–100 g (2.5–3.5 oz) and measuring up to 11 cm (4.3 in). Adult elephant beetles, *Megasoma elephas* and *Megasoma actaeon* often reach 50 g (1.8 oz) and 10 cm (3.9 in).

The longest beetle is the Hercules beetle *Dynastes hercules*, with a maximum overall length of at least 16.7 cm (6.6 in) including the very long pronotal horn. The smallest recorded beetle and the smallest free-living insect (as of 2015), is the featherwing beetle *Scydosella musawasensis* which may measure as little as 325 μm in length.

Titan beetle, *Titanus giganteus*, a tropical longhorn, is one of the largest and heaviest insects in the world.

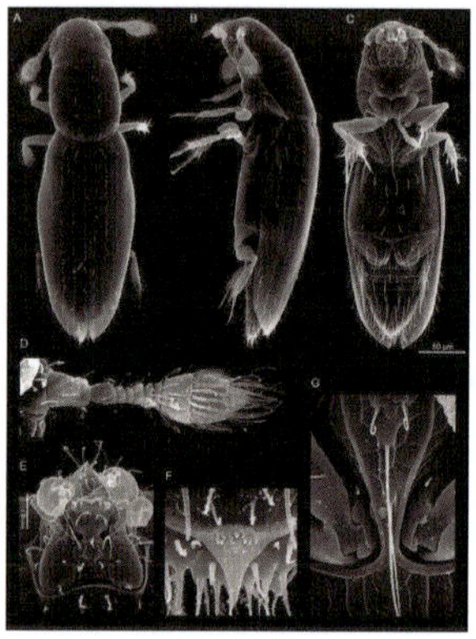

Scydosella musawasensis, the smallest known beetle: scale bar (right) is 50 μm.

Hercules beetle, *Dynastes hercules ecuatorianus*, the longest of all beetles.

Evolution

Late Paleozoic

The oldest known fossil insect that unequivocally resembles a Coleopteran is from the Lower Permian Period about 270 million years ago (mya), though these members of the family Tshekardocoleidae have 13-segmented antennae, elytra with more fully developed venation and more irregular longitudinal ribbing, and abdomen and ovipositor extending beyond the apex of the elytra. In the Permian–Triassic extinction event at the end of the Permian, some 30% of all insect species became extinct, so the fossil record of insects only includes beetles from the Lower Triassic 220 mya. Around this time, during the Late Triassic, fungus-feeding species such as Cupedidae appear in the fossil record. In the stages of the Upper Triassic, alga-feeding insects such as Triaplidae and Hydrophilidae begin to appear, alongside predatory water beetles. The first weevils, including the Obrienidae, appear alongside the first rove beetles (Staphylinidae), which closely resemble recent species. Some entomologists are sceptical that such early insects are so closely related to present-day species, arguing that this is extremely unlikely; for example, the structure of the metepisternum suggests that the Obrienidae could be Archostemata, not weevils at all, despite fossils with weevil-like snouts.

In 2009, a fossil beetle was described from the Pennsylvanian of Mazon Creek, Illinois, pushing the origin of the beetles to an earlier date, 318 to 299 mya. Fossils from this time have been found in Asia and Europe, for instance in the red slate fossil beds of Niedermoschel near Mainz, Germany. Further fossils have been found in Obora, Czech Republic and Tshekarda in the Ural mountains, Russia. However, there are only a few fossils from North America before the middle Permian, although both Asia and North America had been united to Euramerica. The first discoveries from North America made in the Wellington formation of Oklahoma were published in 2005 and 2008.

As a consequence of the Permian–Triassic extinction event, the fossil record of insects is scant, including beetles from the Lower Triassic. However, there are a few exceptions, such as in Eastern Europe. At the Babiy Kamen site in the Kuznetsk Basin, numerous beetle fossils were discovered, including entire specimens of the infraorders Archostemata (e.g. Ademosynidae, Schizocoleidae), Adephaga (e.g., Triaplidae, Trachypachidae) and Polyphaga (e.g. Hydrophilidae, Byrrhidae, Elateroidea). However, species from the families Cupedidae and Schizophoroidae are not present at this site, whereas they dominate at other fossil sites from the Lower Triassic such as Khey-Yaga, Russia, in the Korotaikha Basin.

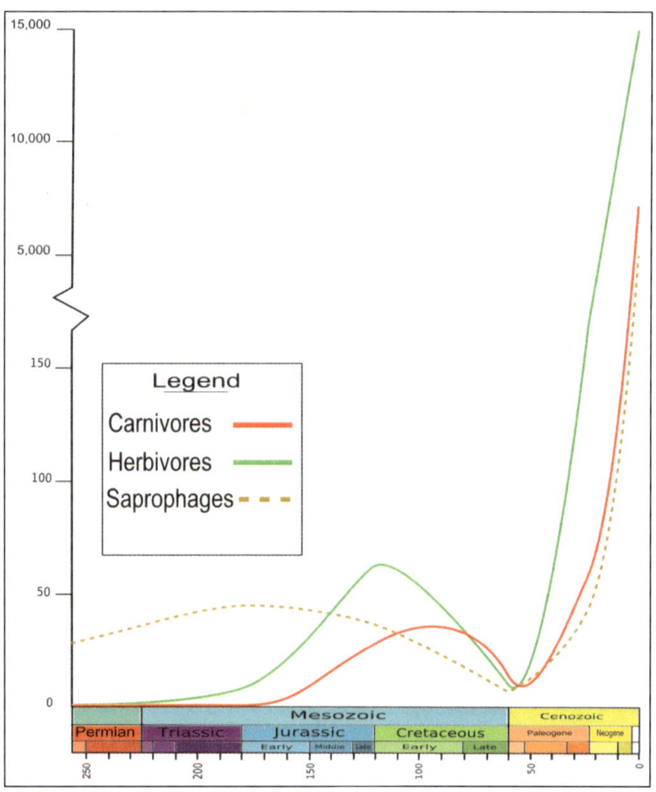

Beetle genera were mainly saprophages (detritivores) in the Permian and Triassic.
During the Jurassic, herbivorous and then carnivorous genera became more common.
In the Cenozoic, genera at all three trophic levels became far more numerous.

Jurassic

During the Jurassic (210 to 145 mya), there was a dramatic increase in the diversity of beetle families, including the development and growth of carnivorous and herbivorous species. The Chrysomeloidea diversified around the same time, feeding on a wide array of plant hosts from cycads and conifers to angiosperms. Close to the Upper Jurassic, the Cupedidae decreased, but the diversity of the early plant-eating species increased. Most recent plant-eating beetles feed on flowering plants or angiosperms, whose success contributed to a doubling of plant-eating species during the Middle Jurassic. However, the increase of the number of beetle families during the Cretaceous does not correlate with the increase of the number of angiosperm species. Around the same time, numerous primitive weevils (e.g. Curculionoidea) and click beetles (e.g. Elateroidea) appeared. The first jewel beetles (e.g. Buprestidae) are present, but they remained rare until the Cretaceous. The first scarab beetles were not coprophagous but presumably fed on rotting wood with the help of fungus; they are an early example of a mutualistic relationship.

There are more than 150 important fossil sites from the Jurassic, the majority in Eastern Europe and North Asia. Outstanding sites include Solnhofen in Upper Bavaria, Germany, Karatau in South Kazakhstan, the Yixian formation in Liaoning, North China, as well as the Jiulongshan formation and further fossil sites in Mongolia. In North America there are only a few sites with fossil records of insects from the Jurassic, namely the shell limestone deposits in the Hartford basin, the Deerfield basin and the Newark basin.

Cretaceous

The Cretaceous saw the fragmenting of the southern landmass, with the opening of the southern Atlantic Ocean and the isolation of New Zealand, while South America, Antarctica, and Australia grew more distant. The diversity of Cupedidae and Archostemata decreased considerably. Predatory ground beetles (Carabidae) and rove beetles (Staphylinidae) began to distribute into different patterns; the Carabidae predominantly occurred in the warm regions, while the Staphylinidae and click beetles (Elateridae) preferred temperate climates. Likewise, predatory species of Cleroidea and Cucujoidea hunted their prey under the bark of trees together with the jewel beetles (Buprestidae). The diversity of jewel beetles increased rapidly, as they were the primary consumers of wood, while longhorn beetles (Cerambycidae) were rather rare: their diversity increased only towards the end of the Upper Cretaceous. The first coprophagous beetles are from the Upper Cretaceous and may have lived on the excrement of herbivorous dinosaurs. The first species where both larvae and adults are adapted to an aquatic lifestyle are found. Whirligig beetles (Gyrinidae) were moderately diverse, although other early beetles (e.g. Dytiscidae) were less, with the most widespread being the species of Coptoclavidae, which preyed on aquatic fly larvae.

Many fossil sites worldwide contain beetles from the Cretaceous. Most are in Europe and Asia and belong to the temperate climate zone during the Cretaceous. Lower Cretaceous sites include the Crato fossil beds in the Araripe basin in the Ceará, North Brazil, as well as overlying Santana formation; the latter was near the equator at that time. In Spain, important sites are near Montsec and Las Hoyas. In Australia, the Koonwarra fossil beds of the Korumburra group, South Gippsland, Victoria, are noteworthy. Major sites from the Upper Cretaceous include Kzyl-Dzhar in South Kazakhstan and Arkagala in Russia.

Cenozoic

Fossil buprestid beetle from the Eocene (50 mya) Messel pit, which retains its structural color.

Beetle fossils are abundant in the Cenozoic; by the Quaternary (up to 1.6 mya), fossil species are identical to living ones, while from the Late Miocene (5.7 mya) the fossils are still so close to modern forms that they are most likely the ancestors of living species. The large oscillations in climate during the Quaternary caused beetles to change their geographic distributions so much that current location gives little clue to the biogeographical history of a species. It is evident that geographic isolation of populations must often have been broken as insects moved under the influence of changing climate, causing mixing of gene pools, rapid evolution, and extinctions, especially in middle latitudes.

External Morphology

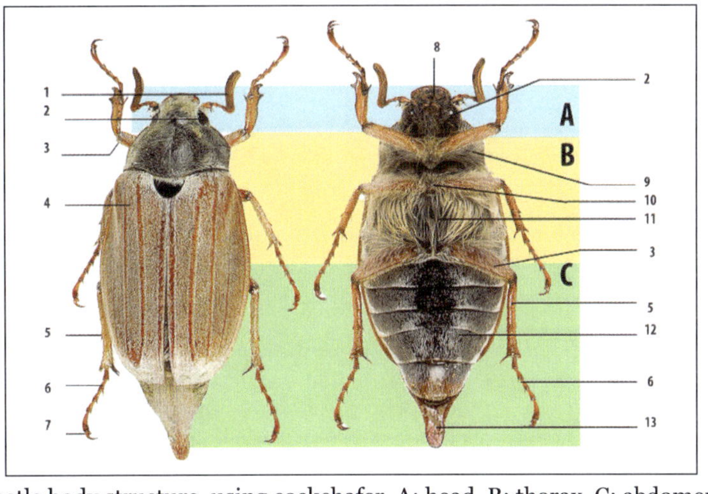

Beetle body structure, using cockchafer. A: head, B: thorax, C: abdomen.
1: antenna, 2: compound eye, 3: femur, 4: elytron (wing cover), 5: tibia, 6: tarsus, 7: claws,
8: mouthparts, 9: prothorax, 10: mesothorax, 11: metathorax, 12: abdominal sternites, 13: pygidium.

Beetles are generally characterized by a particularly hard exoskeleton and hard forewings (elytra) not usable for flying. Almost all beetles have mandibles that move in a horizontal plane. The mouthparts are rarely suctorial, though they are sometimes reduced; the maxillae always bear palps. The antennae usually have 11 or fewer segments, except in some groups like the Cerambycidae (longhorn beetles) and the Rhipiceridae (cicada parasite beetles). The coxae of the legs are usually located recessed within a coxal cavity. The genitalic structures are telescoped into the last abdominal segment in all extant beetles. Beetle larvae can often be confused with those of other endopterygote groups. The beetle's exoskeleton is made up of numerous plates, called sclerites, separated by thin sutures. This design provides armored defenses while maintaining flexibility. The general anatomy of a beetle is quite uniform, although specific organs and appendages vary greatly in appearance and function between the many families in the order. Like all insects, beetles' bodies are divided into three sections: the head, the thorax, and the abdomen. Because there are so many species, identification is quite difficult, and relies on attributes including the shape of the antennae, the tarsal formulae and shapes of these small segments on the legs, the mouthparts, and the ventral plates (sterna, pleura, coxae). In many species accurate identification can only be made by examination of the unique male genitalic structures.

Head

The head, having mouthparts projecting forward or sometimes downturned, is usually heavily sclerotized and is sometimes very large. The eyes are compound and may display remarkable adaptability, as in the case of the aquatic whirligig beetles (Gyrinidae), where they are split to allow a view both above and below the waterline. A few Longhorn beetles (Cerambycidae) and weevils as well as some fireflies (Rhagophthalmidae) have divided eyes, while many have eyes that are notched, and a few have ocelli, small, simple eyes usually farther back on the head (on the vertex); these are more common in larvae than in adults. The anatomical organization of the compound eyes may be modified and depends on whether a species is primarily crepuscular, or diurnally or nocturnally active. Ocelli are found in the adult carpet beetle (Dermestidae), some rove beetles (Omaliinae), and the Derodontidae.

Polyphylla fullo has distinctive fan-like antennae, one of several distinct forms for the appendages among beetles.

Beetle antennae are primarily organs of sensory perception and can detect motion, odour and chemical substances, but may also be used to physically feel a beetle's environment. Beetle families may use antennae in different ways. For example, when moving quickly, tiger beetles may not be able to see very well and instead hold their antennae rigidly in front of them in order to avoid obstacles. Certain Cerambycidae use antennae to balance, and blister beetles may use them for grasping. Some aquatic beetle species may use antennae for gathering air and passing it under the body whilst submerged. Equally, some families use antennae during mating, and a few species use them for defence. In the cerambycid *Onychocerus albitarsis*, the antennae have venom injecting structures used in defence, which is unique among arthropods. Antennae vary greatly in form, sometimes between the sexes, but are often similar within any given family. Antennae may be clubbed, threadlike, angled, shaped like a string of beads, comb-like (either on one side or both, bipectinate), or toothed. The physical variation of antennae is important for the identification of many beetle groups. The Curculionidae have elbowed or geniculate antennae. Feather like flabellate antennae are a restricted form found in the Rhipiceridae and a few other families. The Silphidae have a capitate antennae with a spherical head at the tip. The Scarabaeidae typically have lamellate antennae with the terminal segments extended into long flat structures stacked together. The Carabidae typically have thread-like antennae. The antennae arises between the eye and the mandibles and in the Tenebrionidae, the antennae rise in front of a notch that breaks the usually circular outline of the compound eye. They are segmented and usually consist of 11 parts, the first part is called the scape and the second part is the pedicel. The other segments are jointly called the flagellum.

Beetles have mouthparts like those of grasshoppers. The mandibles appear as large pincers on the front of some beetles. The mandibles are a pair of hard, often tooth-like structures that move horizontally to grasp, crush, or cut food or enemies. Two pairs of finger-like appendages, the maxillary and labial palpi, are found around the mouth in most beetles, serving to move food into the mouth. In many species, the mandibles are sexually dimorphic, with those of the males enlarged enormously compared with those of females of the same species.

Thorax

The thorax is segmented into the two discernible parts, the pro- and pterothorax. The pterothorax is the fused meso- and metathorax, which are commonly separated in other insect species, although flexibly articulate from the prothorax. When viewed from below, the thorax is that part from which all three pairs of legs and both pairs of wings arise. The abdomen is everything posterior to the thorax. When viewed from above, most beetles appear to have three clear sections, but this is deceptive: on the beetle's upper surface, the middle section is a hard plate called the pronotum, which is only the front part of the thorax; the back part of the thorax is concealed by the beetle's wings. This further segmentation is usually best seen on the abdomen.

Acilius sulcatus, a diving beetle with hind legs adapted as swimming limbs.

Legs

The multisegmented legs end in two to five small segments called tarsi. Like many other insect orders, beetles have claws, usually one pair, on the end of the last tarsal segment of each leg. While most beetles use their legs for walking, legs have been variously adapted for other uses. Aquatic beetles including the Dytiscidae (diving beetles), Haliplidae, and many species of Hydrophilidae, the legs, often the last pair, are modified for swimming, typically with rows of long hairs. Male diving beetles have suctorial cups on their forelegs that they use to grasp females. Other beetles have fossorial legs widened and often spined for digging. Species with such adaptations are found among the scarabs, ground beetles, and clown beetles (Histeridae). The hind legs of some beetles, such as flea beetles (within Chrysomelidae) and flea weevils (within Curculionidae), have enlarged femurs that help them leap.

Wings

Checkered beetle *Trichodes alvearius* taking off, showing the hard elytra (forewings adapted as wing-cases) held stiffly away from the flight wings.

The forewings of beetles are not used for flight, but form elytra which cover the hind part of the body and protect the hindwings. The elytra are usually hard shell-like structures which must be raised to allow the hind wings to move for flight. However, in the soldier beetles (Cantharidae), the elytra are soft, earning this family the name of leatherwings. Other soft wing beetles include the net-winged beetle *Calopteron discrepans*, which has brittle wings that rupture easily in order to release chemicals for defence.

Beetles' flight wings are crossed with veins and are folded after landing, often along these veins, and stored below the elytra. A fold (*jugum*) of the membrane at the base of each wing is characteristic. Some beetles have lost the ability to fly. These include some ground beetles (Carabidae) and some true weevils (Curculionidae), as well as desert- and cave-dwelling species of other families. Many have the two elytra fused together, forming a solid shield over the abdomen. In a few families, both the ability to fly and the elytra have been lost, as in the glow-worms (Phengodidae), where the females resemble larvae throughout their lives. The presence of elytra and wings does not always indicate that the beetle will fly. For example, the tansy beetle walks between habitats despite being physically capable of flight.

Abdomen

The abdomen is the section behind the metathorax, made up of a series of rings, each with a hole for breathing and respiration, called a spiracle, composing three different segmented sclerites: the tergum, pleura, and the sternum. The tergum in almost all species is membranous, or usually soft and concealed by the wings and elytra when not in flight. The pleura are usually small or hidden in some species, with each pleuron having a single spiracle. The sternum is the most widely visible part of the abdomen, being a more or less sclerotized segment. The abdomen itself does not have any appendages, but some (for example, Mordellidae) have articulating sternal lobes.

Anatomy and Physiology

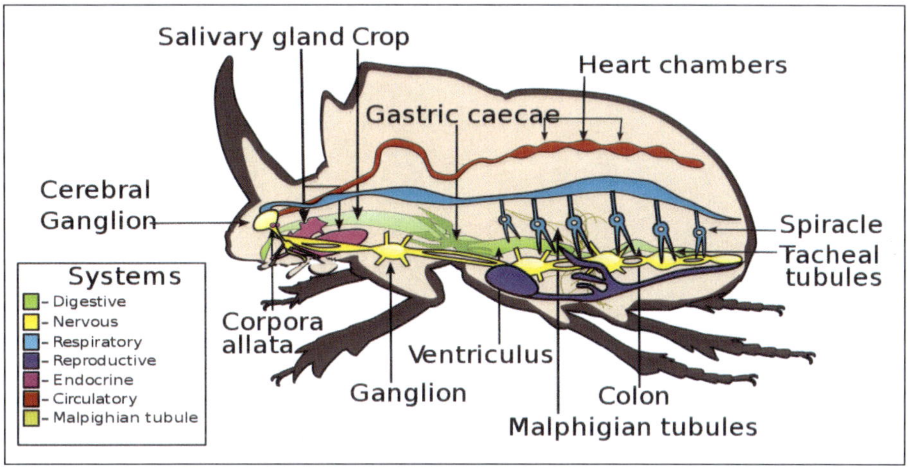

A beetle's body systems.

Digestive System

The digestive system of beetles is primarily adapted for a herbivorous diet. Digestion takes place mostly in the anterior midgut, although in predatory groups like the Carabidae, most digestion occurs in the crop by means of midgut enzymes. In the Elateridae, the larvae are liquid feeders that extraorally digest their food by secreting enzymes. The alimentary canal basically consists of a short, narrow pharynx, a widened expansion, the crop, and a poorly developed gizzard. This is followed by the midgut, that varies in dimensions between species, with a large amount of cecum, and the hindgut, with varying lengths. There are typically four to six Malpighian tubules.

Nervous System

The nervous system in beetles contains all the types found in insects, varying between different species, from three thoracic and seven or eight abdominal ganglia which can be distinguished to that in which all the thoracic and abdominal ganglia are fused to form a composite structure.

Respiratory System

Like most insects, beetles inhale air, for the oxygen it contains, and exhale carbon dioxide, via a tracheal system. Air enters the body through spiracles, and circulates within the haemocoel in a system of tracheae and tracheoles, through whose walls the gases can diffuse.

Diving beetles, such as the Dytiscidae, carry a bubble of air with them when they dive. Such a bubble may be contained under the elytra or against the body by specialized hydrophobic hairs. The bubble covers at least some of the spiracles, permitting air to enter the tracheae. The function of the bubble is not only to contain a store of air but to act as a physical gill. The air that it traps is in contact with oxygenated water, so as the animal's consumption depletes the oxygen in the bubble, more oxygen can diffuse in to replenish it. Carbon dioxide is more soluble in water than either oxygen or nitrogen, so it readily diffuses out faster than in. Nitrogen is the most plentiful gas in the bubble, and the least soluble, so it constitutes a relatively static component of the bubble and acts as a stable medium for respiratory gases to accumulate in and pass through. Occasional visits to the surface are sufficient for the beetle to re-establish the constitution of the bubble.

Dytiscus spiracles (right) on upper side of abdomen, normally covered by the elytra, are in contact with an air bubble when the beetle dives.

Circulatory System

Like other insects, beetles have open circulatory systems, based on hemolymph rather than blood. As in other insects, a segmented tube-like heart is attached to the dorsal wall of the hemocoel. It has paired inlets or *ostia* at intervals down its length, and circulates the hemolymph from the main cavity of the haemocoel and out through the anterior cavity in the head.

Specialized Organs

Different glands are specialized for different pheromones to attract mates. Pheromones from species of Rutelinae are produced from epithelial cells lining the inner surface of the apical abdominal segments; amino acid-based pheromones of Melolonthinae are produced from eversible glands on the abdominal apex. Other species produce different types of pheromones. Dermestids produce esters, and species of Elateridae produce fatty acid-derived aldehydes and acetates. To attract a mate, fireflies (Lampyridae) use modified fat body cells with transparent surfaces backed with reflective uric acid crystals to produce light by bioluminescence. Light production is highly efficient, by oxidation of luciferin catalyzed by enzymes (luciferases) in the presence of adenosine triphosphate (ATP) and oxygen, producing oxyluciferin, carbon dioxide, and light.

Tympanal organs or hearing organs consist of a membrane (tympanum) stretched across a frame backed by an air sac and associated sensory neurons, are found in two families. Several species of the genus *Cicindela* (Carabidae) have hearing organs on the dorsal surfaces of their first abdominal segments beneath the wings; two tribes in the Dynastinae (within the Scarabaeidae) have hearing organs just beneath their pronotal shields or neck membranes. Both families are sensitive to ultrasonic frequencies, with strong evidence indicating they function to detect the presence of bats by their ultrasonic echolocation.

Reproduction and Development

Beetles are members of the superorder Endopterygota, and accordingly most of them undergo complete metamorphosis. The typical form of metamorphosis in beetles passes through four main stages: the egg, the larva, the pupa, and the imago or adult. The larvae are commonly called grubs and the pupa sometimes is called the chrysalis. In some species, the pupa may be enclosed in a cocoon constructed by the larva towards the end of its final instar. Some beetles, such as typical members of the families Meloidae and Rhipiphoridae, go further, undergoing hypermetamorphosis in which the first instar takes the form of a triungulin.

Mating

Punctate flower chafers (*Neorrhina punctata*, Scarabaeidae) mating.

Some beetles have intricate mating behavior. Pheromone communication is often important in locating a mate. Different species use different pheromones. Scarab beetles such as the Rutelinae use pheromones derived from fatty acid synthesis, while other scarabs such as the Melolonthinae use amino acids and terpenoids. Another way beetles find mates is seen in the fireflies (Lampyridae) which are bioluminescent, with abdominal light-producing organs. The males and females engage in a complex dialogue before mating; each species has a unique combination of flight patterns, duration, composition, and intensity of the light produced.

Before mating, males and females may stridulate, or vibrate the objects they are on. In the Meloidae, the male climbs onto the dorsum of the female and strokes his antennae on her head, palps, and antennae. In *Eupompha*, the male draws his antennae along his longitudinal vertex. They may not mate at all if they do not perform the precopulatory ritual. This mating behavior may be different amongst dispersed populations of the same species. For example, the mating of a Russian population of tansy beetle (*Chysolina graminis*) is preceded by an elaborate ritual involving the male tapping the female's eyes, pronotum and antennae with its antennae, which is not evident in the population of this species in the United Kingdom.

Competition can play a part in the mating rituals of species such as burying beetles (*Nicrophorus*), the insects fighting to determine which can mate. Many male beetles are territorial and fiercely defend their territories from intruding males. In such species, the male often has horns on the head or thorax, making its body length greater than that of a female. Copulation is generally quick, but in some cases lasts for several hours. During copulation, sperm cells are transferred to the female to fertilize the egg.

Life Cycle

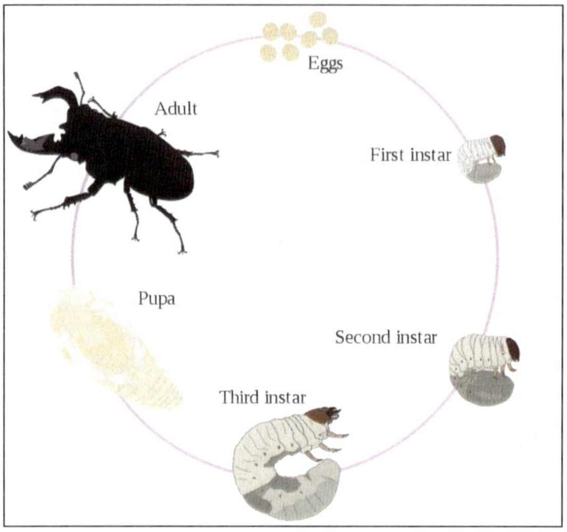

The life cycle of the stag beetle includes three instars.

Egg

Essentially all beetles lay eggs, though some myrmecophilous Aleocharinae and some Chrysomelinae which live in mountains or the subarctic are ovoviviparous, laying eggs which hatch almost immediately. Beetle eggs generally have smooth surfaces and are soft, though the Cupedidae have hard eggs. Eggs vary widely between species: the eggs tend to be small in species with many instars (larval stages), and in those that lay large numbers of eggs. A female may lay from several dozen to several thousand eggs during her lifetime, depending on the extent of parental care. This ranges from the simple laying of eggs under a leaf, to the parental care provided by scarab beetles, which house, feed and protect their young. The Attelabidae roll leaves and lay their eggs inside the roll for protection.

Larva

The larva is usually the principal feeding stage of the beetle life cycle. Larvae tend to feed voraciously once they emerge from their eggs. Some feed externally on plants, such as those of certain leaf beetles, while others feed within their food sources. Examples of internal feeders are most Buprestidae and longhorn beetles. The larvae of many beetle families are predatory like the adults (ground beetles, ladybirds, rove beetles). The larval period varies between species, but can be as long as several years. The larvae of skin beetles undergo a degree of reversed development when starved, and later grow back to the previously attained level of maturity. The cycle can be repeated many times. Larval morphology is highly varied amongst species, with well-developed and sclerotized heads, distinguishable thoracic and abdominal segments (usually the tenth, though sometimes the eighth or ninth).

Scarabaeiform larva of Hercules beetle.

Beetle larvae can be differentiated from other insect larvae by their hardened, often darkened heads, the presence of chewing mouthparts, and spiracles along the sides of their bodies. Like adult beetles, the larvae are varied in appearance, particularly between beetle families. Beetles with somewhat flattened, highly mobile larvae include the ground beetles and rove beetles; their larvae are described as campodeiform. Some beetle larvae resemble hardened worms with dark head capsules and minute legs. These are elateriform larvae, and are found in the click beetle (Elateridae) and darkling beetle (Tenebrionidae) families. Some elateriform larvae of click beetles are known as wireworms. Beetles in the Scarabaeoidea have short, thick larvae described as scarabaeiform, more commonly known as grubs.

All beetle larvae go through several instars, which are the developmental stages between each moult. In many species, the larvae simply increase in size with each successive instar as more food is consumed. In some cases, however, more dramatic changes occur. Among certain beetle families or genera, particularly those that exhibit parasitic lifestyles, the first instar (the planidium) is highly mobile to search out a host, while the following instars are more sedentary and remain on or within their host. This is known as hypermetamorphosis; it occurs in the Meloidae, Micromalthidae, and Ripiphoridae. The blister beetle *Epicauta vittata* (Meloidae), for example, has three distinct larval stages. Its first stage, the triungulin, has longer legs to go in search of the eggs of grasshoppers. After feeding for a week it moults to the second stage, called the caraboid stage, which resembles the larva of a carabid beetle. In another week it moults and assumes the appearance of a scarabaeid larva – the scarabaeidoid stage. Its penultimate larval stage is the pseudo-pupa or the coarcate larva, which will overwinter and pupate until the next spring.

The larval period can vary widely. A fungus feeding staphylinid *Phanerota fasciata* undergoes three moults in 3.2 days at room temperature while *Anisotoma* sp. (Leiodidae) completes its larval stage in the fruiting body of slime mold in 2 days and possibly represents the fastest growing beetles. Dermestid beetles, *Trogoderma inclusum* can remain in an extended larval state under unfavourable conditions, even reducing their size between moults. A larva is reported to have survived for 3.5 years in an enclosed container.

Pupa and Adult

The ivory-marked beetle, *Eburia quadrigeminata*, may live up to 40 years inside the hardwoods on which the larva feeds.

As with all endopterygotes, beetle larvae pupate, and from these pupae emerge fully formed, sexually mature adult beetles, or imagos. Pupae never have mandibles (they are adecticous). In most pupae, the appendages are not attached to the body and are said to be exarate; in a few beetles (Staphylinidae, Ptiliidae etc.) the appendages are fused with the body (termed as obtect pupae).

Adults have extremely variable lifespans, from weeks to years, depending on the species. Some wood-boring beetles can have extremely long life-cycles. It is believed that when furniture or house timbers are infested by beetle larvae, the timber already contained the larvae when it was first sawn up. A birch bookcase 40 years old released adult *Eburia quadrigeminata* (Cerambycidae), while *Buprestis aurulenta* and other Buprestidae have been documented as emerging as much as 51 years after manufacture of wooden items.

Behavior

Locomotion

Photinus pyralis, firefly, in flight.

The elytra allow beetles to both fly and move through confined spaces, doing so by folding the delicate wings under the elytra while not flying, and folding their wings out just before takeoff. The unfolding and folding of the wings is operated by muscles attached to the wing base; as long as the tension on the radial and cubital veins remains, the wings remain straight. In some day-flying species (for example, Buprestidae, Scarabaeidae), flight does not include large amounts of lifting of the elytra, having the metathorac wings extended under the lateral elytra margins. The altitude reached by beetles in flight varies. One study investigating the flight altitude of the ladybird species *Coccinella septempunctata* and *Harmonia axyridis* using radar showed that, whilst the majority in flight over a single location were at 150–195 m above ground level, some reached altitudes of over 1100 m.

Many rove beetles have greatly reduced elytra, and while they are capable of flight, they most often move on the ground: their soft bodies and strong abdominal muscles make them flexible, easily able to wriggle into small cracks.

Aquatic beetles use several techniques for retaining air beneath the water's surface. Diving beetles (Dytiscidae) hold air between the abdomen and the elytra when diving. Hydrophilidae have hairs on their under surface that retain a layer of air against their bodies. Adult crawling water beetles use both their elytra and their hind coxae (the basal segment of the back legs) in air retention, while whirligig beetles simply carry an air bubble down with them whenever they dive.

Communication

Beetles have a variety of ways to communicate, including the use of pheromones. The mountain pine beetle emits a pheromone to attract other beetles to a tree. The mass of beetles are able to overcome the chemical defenses of the tree. After the tree's defenses have been exhausted, the beetles emit an anti-aggregation pheromone. This species can stridulate to communicate, but others may use sound to defend themselves when attacked.

Parental Care

A dung beetle rolling dung.

Parental care is found in a few families of beetle, perhaps for protection against adverse conditions and predators. The rove beetle *Bledius spectabilis* lives in salt marshes, so the eggs and larvae are endangered by the rising tide. The maternal beetle patrols the eggs and larvae, burrowing to keep them from flooding and asphyxiating, and protects them from the predatory carabid beetle *Dicheirotrichus gustavi* and from the parasitoidal wasp *Barycnemis blediator*, which kills some 15% of the larvae.

Burying beetles are attentive parents, and participate in cooperative care and feeding of their offspring. Both parents work to bury small animal carcass to serve as a food resource for their young and build a brood chamber around it. The parents prepare the carcass and protect it from competitors and from early decomposition. After their eggs hatch, the parents keep the larvae clean of fungus and bacteria and help the larvae feed by regurgitating food for them.

Some dung beetles provide parental care, collecting herbivore dung and laying eggs within that food supply, an instance of mass provisioning. Some species do not leave after this stage, but remain to safeguard their offspring.

Most species of beetles do not display parental care behaviors after the eggs have been laid.

Subsociality, where females guard their offspring, is well-documented in two families of Chrysomelidae, Cassidinae and Chrysomelinae.

Eusociality

Eusociality involves cooperative brood care (including brood care of offspring from other individuals), overlapping generations within a colony of adults, and a division of labour into reproductive and non-reproductive groups. Few organisms outside Hymenoptera exhibit this behavior; the only beetle to do so is the weevil *Austroplatypus incompertus*. This Australian species lives in horizontal networks of tunnels, in the heartwood of *Eucalyptus* trees. It is one of more than 300 species of wood-boring Ambrosia beetles which distribute the spores of ambrosia fungi. The fungi grow in the beetles' tunnels, providing food for the beetles and their larvae; female offspring remain in the tunnels and maintain the fungal growth, probably never reproducing. Cooperative brood care is also found in the bess beetles (Passalidae) where the larvae feed on the semi-digested faeces of the adults.

Feeding

Hycleus sp. (Meloidae) feeding on the petals of *Ipomoea carnea*.

Beetles are able to exploit a wide diversity of food sources available in their many habitats. Some are omnivores, eating both plants and animals. Other beetles are highly specialized in their diet. Many species of leaf beetles, longhorn beetles, and weevils are very host-specific, feeding on only a single species of plant. Ground beetles and rove beetles (Staphylinidae), among others, are primarily carnivorous and catch and consume many other arthropods and small prey, such as earthworms and snails. While most predatory beetles are generalists, a few species have more specific prey requirements or preferences.

Decaying organic matter is a primary diet for many species. This can range from dung, which is consumed by coprophagous species (such as certain scarab beetles in the Scarabaeidae), to dead animals, which are eaten by necrophagous species (such as the carrion beetles, Silphidae). Some beetles found in dung and carrion are in fact predatory. These include members of the Histeridae and Silphidae, preying on the larvae of coprophagous and necrophagous insects. Many beetles feed under bark, some feed on wood while others feed on fungi growing on wood or leaf-litter. Some beetles have special mycangia, structures for the transport of fungal spores.

Ecology

A camouflaged longhorn beetle, *Ecyrus dasycerus.*

Anti-predator Adaptations

Beetles, both adults and larvae, are the prey of many animal predators including mammals from bats to rodents, birds, lizards, amphibians, fishes, dragonflies, robberflies, reduviid bugs, ants, other beetles, and spiders. Beetles use a variety of anti-predator adaptations to defend themselves. These include camouflage and mimicry against predators that hunt by sight, toxicity, and defensive behavior.

Camouflage

Camouflage is common and widespread among beetle families, especially those that feed on wood or vegetation, such as leaf beetles (Chrysomelidae, which are often green) and weevils. In some species, sculpturing or various coloured scales or hairs cause beetles such as the avocado weevil *Heilipus apiatus* to resemble bird dung or other inedible objects. Many beetles that live in sandy environments blend in with the coloration of that substrate.

Mimicry and Aposematism

Clytus arietis (Cerambycidae), a Batesian mimic of wasps.

Some longhorn beetles (Cerambycidae) are effective Batesian mimics of wasps. Beetles may combine coloration with behavioural mimicry, acting like the wasps they already closely resemble. Many other beetles, including ladybirds, blister beetles, and lycid beetles secrete distasteful or toxic substances to make them unpalatable or poisonous, and are often aposematic, where bright or contrasting coloration warn off predators; many beetles and other insects mimic these chemically protected species.

Blister beetles such as *Hycleus* have brilliant aposematic coloration, warning of their toxicity.

The bloody-nosed beetle, *Timarcha tenebricosa*, defending itself by releasing a droplet of noxious red liquid (base of leg, on right).

Chemical defense is important in some species, usually being advertised by bright aposematic colours. Some Tenebrionidae use their posture for releasing noxious chemicals to warn off predators.

Chemical defences may serve purposes other than just protection from vertebrates, such as protection from a wide range of microbes. Some species sequester chemicals from the plants they feed on, incorporating them into their own defenses.

Other species have special glands to produce deterrent chemicals. The defensive glands of carabid ground beetles produce a variety of hydrocarbons, aldehydes, phenols, quinones, esters, and acids released from an opening at the end of the abdomen. African carabid beetles (for example, *Anthia* and *Thermophilum* – *Thermophilum* is sometimes included within *Anthia*) employ the same chemicals as ants: formic acid. Bombardier beetles have well-developed pygidial glands that empty from the sides of the intersegment membranes between the seventh and eighth abdominal segments. The gland is made of two containing chambers, one for hydroquinones and hydrogen peroxide, the other holding hydrogen peroxide and catalase enzymes. These chemicals mix and result in an explosive ejection, reaching a temperature of around 100 °C (212 °F), with the breakdown of hydroquinone to hydrogen, oxygen, and quinone. The oxygen propels the noxious chemical spray as a jet that can be aimed accurately at predators.

Other Defences

Large ground-dwelling beetles such as Carabidae, the rhinoceros beetle and the longhorn beetles defend themselves using strong mandibles, or heavily sclerotised (armored) spines or horns to deter or fight off predators. Many species of weevil that feed out in the open on leaves of plants react to attack by employing a drop-off reflex. Some combine it with thanatosis, in which they close up their appendages and "play dead". The click beetles (Elateridae) can suddenly catapult themselves out of danger by releasing the energy stored by a click mechanism, which consists of a stout spine on the prosternum and a matching groove in the mesosternum. Some species startle an attacker by producing sounds through a process known as stridulation.

Parasitism

A few species of beetles are ectoparasitic on mammals. One such species, *Platypsyllus castoris*, parasitises beavers (*Castor* spp.). This beetle lives as a parasite both as a larva and as an adult, feeding on epidermal tissue and possibly on skin secretions and wound exudates. They are strikingly flattened dorsoventrally, no doubt as an adaptation for slipping between the beavers' hairs. They are wingless and eyeless, as are many other ectoparasites. Others are kleptoparasites of other invertebrates, such as the small hive beetle (*Aethina tumida*) that infests honey bee nests, while many species are parasitic inquilines or commensal in the nests of ants. A few groups of beetles are primary parasitoids of other insects, feeding off of, and eventually killing their hosts.

Pollination

Beetle-pollinated flowers are usually large, greenish or off-white in color, and heavily scented. Scents may be spicy, fruity, or similar to decaying organic material. Beetles were most likely the first insects to pollinate flowers. Most beetle-pollinated flowers are flattened or dish-shaped, with pollen easily accessible, although they may include traps to keep the beetle longer. The plants' ovaries are usually well protected from the biting mouthparts of their pollinators. The beetle families that habitually pollinate flowers are the Buprestidae, Cantharidae, Cerambycidae, Cleridae, Dermestidae, Lycidae, Melyridae, Mordellidae, Nitidulidae and Scarabaeidae. Beetles may be particularly important in some parts of the world such as semiarid areas of southern Africa and southern California and the montane grasslands of KwaZulu-Natal in South Africa.

An Israeli Copper Flower-Chafer (*Protaetia cuprea ignicollis*) pollinating a crown daisy (*Glebionis coronaria*).

Mutualism

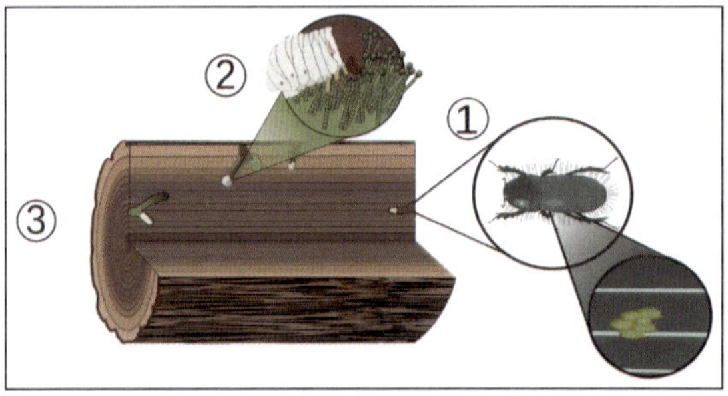

1: Adult ambrosia beetle burrows into wood and lays eggs, carrying fungal spores in its mycangia. 2: Larva feeds on fungus, which digests wood, removing toxins, to mutual benefit. 3: Larva pupates.

Mutualism is well known in a few beetles, such as the ambrosia beetle, which partners with fungi to digest the wood of dead trees. The beetles excavate tunnels in dead trees in which they cultivate fungal gardens, their sole source of nutrition. After landing on a suitable tree, an ambrosia beetle excavates a tunnel in which it releases spores of its fungal symbiont. The fungus penetrates the plant's xylem tissue, digests it, and concentrates the nutrients on and near the surface of the beetle gallery, so the weevils and the fungus both benefit. The beetles cannot eat the wood due to toxins, and uses its relationship with fungi to help overcome the defenses of its host tree in order to provide nutrition for their larvae. Chemically mediated by a bacterially produced polyunsaturated peroxide, this mutualistic relationship between the beetle and the fungus is coevolved.

Beetle found in Thar Desert.

Tolerance of Extreme Environments

The fogstand beetle of the Namib Desert, *Stenocara gracilipes* is able to survive by collecting water from fog on its back.

About 90% of beetle species enter a period of adult diapause, a quiet phase with reduced metabolism to tide unfavourable environmental conditions. Adult diapause is the most common form of diapause in Coleoptera. To endure the period without food (often lasting many months) adults prepare by accumulating reserves of lipids, glycogen, proteins and other substances needed for resistance to future hazardous changes of environmental conditions. This diapause is induced by signals heralding the arrival of the unfavourable season; usually the cue is photoperiodic. Short (decreasing) day length serves as a signal of approaching winter and induces winter diapause (hibernation). A study of hibernation in the Arctic beetle *Pterostichus brevicorni* showed that the body fat levels of adults were highest in autumn with the alimentary canal filled with food, but empty by the end of January. This loss of body fat was a gradual process, occurring in combination with dehydration.

All insects are poikilothermic, so the ability of a few beetles to live in extreme environments depends on their resilience to unusually high or low temperatures. The bark beetle *Pityogenes chalcographus* can survive −39 °C whilst overwintering beneath tree bark; the Alaskan beetle *Cucujus clavipes puniceus* is able to withstand −58 °C; its larvae may survive −100 °C. At these low temperatures, the formation of ice crystals in internal fluids is the biggest threat to survival to beetles, but this is prevented through the production of antifreeze proteins that stop water molecules from grouping together. The low temperatures experienced by *Cucujus clavipes* can be survived through their deliberate dehydration in conjunction with the antifreeze proteins. This concentrates the antifreezes several fold. The hemolymph of the mealworm beetle *Tenebrio molitor* contains several antifreeze proteins. The Alaskan beetle *Upis ceramboides* can survive −60 °C: its cryoprotectants are xylomannan, a molecule consisting of a sugar bound to a fatty acid, and the sugar-alcohol, threitol.

Conversely, desert dwelling beetles are adapted to tolerate high temperatures. For example, the Tenebrionid beetle *Onymacris rugatipennis* can withstand 50 °C. Tiger beetles in hot, sandy areas are often whitish (for example, *Habroscelimorpha dorsalis*), to reflect more heat than a darker colour would. These beetles also exhibits behavioural adaptions to tolerate the heat: they are able to stand erect on their tarsi to hold their bodies away from the hot ground, seek shade, and turn to face the sun so that only the front parts of their heads are directly exposed.

The fogstand beetle of the Namib Desert, *Stenocara gracilipes*, is able to collect water from fog, as its elytra have a textured surface combining hydrophilic (water-loving) bumps and waxy, hydrophobic troughs. The beetle faces the early morning breeze, holding up its abdomen; droplets condense on the elytra and run along ridges towards their mouthparts. Similar adaptations are found in several other Namib desert beetles such as *Onymacris unguicularis*.

Some terrestrial beetles that exploit shoreline and floodplain habitats have physiological adaptations for surviving floods. In the event of flooding, adult beetles may be mobile enough to move away from flooding, but larvae and pupa often cannot. Adults of *Cicindela togata* are unable to survive immersion in water, but larvae are able to survive a prolonged period, up to 6 days, of anoxia during floods. Anoxia tolerance in the larvae may have been sustained by switching to anaerobic metabolic pathways or by reducing metabolic rate. Anoxia tolerance in the adult Carabid beetle *Pelophilia borealis* was tested in laboratory conditions and it was found that they could survive a continuous period of up to 127 days in an atmosphere of 99.9% nitrogen at 0 °C.

Migration

Many beetle species undertake annual mass movements which are termed as migrations. These include the pollen beetle *Meligethes aeneus* and many species of coccinellids. These mass movements may also be opportunistic, in search of food, rather than seasonal. A 2008 study of an unusually large outbreak of Mountain Pine Beetle (*Dendroctonus ponderosae*) in British Columbia found that beetles were capable of flying 30–110 km per day in densities of up to 18, 600 beetles per hectare.

Myrmecology

Meat eater ant feeding on honey.

Myrmecology is a branch of entomology focusing on the scientific study of ants. Some early myrmecologists considered ant society as the ideal form of society and sought to find solutions to human problems by studying them. Ants continue to be a model of choice for the study of questions on the evolution of social systems because of their complex and varied forms of eusociality (social organization). Their diversity and prominence in ecosystems also has made them important components in the study of biodiversity and conservation. Recently, ant colonies are also studied and modeled for their relevance in machine learning, complex interactive networks, stochasticity of encounter and interaction networks, parallel computing, and other computing fields.

Ants

Ants are eusocial insects of the family Formicidae and, along with the related wasps and bees, belong to the order Hymenoptera. Ants evolved from wasp-like ancestors in the Cretaceous period, about 140 million years ago, and diversified after the rise of flowering plants. More than 12,500 of an estimated total of 22,000 species have been classified. They are easily identified by their elbowed antennae and the distinctive node-like structure that forms their slender waists.

Ants form colonies that range in size from a few dozen predatory individuals living in small natural cavities to highly organised colonies that may occupy large territories and consist of millions of individuals. Larger colonies consist of various castes of sterile, wingless females, most of which are workers (ergates), as well as soldiers (dinergates) and other specialised groups. Nearly all ant colonies also have some fertile males called "drones" (aner) and one or more fertile females called "queens" (gynes). The colonies are described as superorganisms because the ants appear to operate as a unified entity, collectively working together to support the colony.

Ants have colonised almost every landmass on Earth. The only places lacking indigenous ants are Antarctica and a few remote or inhospitable islands. Ants thrive in most ecosystems and may form 15–25% of the terrestrial animal biomass. Their success in so many environments has been attributed to their social organisation and their ability to modify habitats, tap resources, and defend themselves. Their long co-evolution with other species has led to mimetic, commensal, parasitic, and mutualistic relationships.

Ant societies have division of labour, communication between individuals, and an ability to solve complex problems. These parallels with human societies have long been an inspiration and subject of study. Many human cultures make use of ants in cuisine, medication, and rituals. Some species are valued in their role as biological pest control agents. Their ability to exploit resources may bring ants into conflict with humans, however, as they can damage crops and invade buildings. Some species, such as the red imported fire ant (*Solenopsis invicta*), are regarded as invasive species, establishing themselves in areas where they have been introduced accidentally.

Taxonomy and Evolution

The family Formicidae belongs to the order Hymenoptera, which also includes sawflies, bees, and wasps. Ants evolved from a lineage within the stinging wasps, and a 2013 study suggests that they are a sister group of the Apoidea. In 1966, E. O. Wilson and his colleagues identified the fossil remains of an ant (*Sphecomyrma*) that lived in the Cretaceous period. The specimen, trapped in amber dating back to around 92 million years ago, has features found in some wasps, but not found in modern ants. *Sphecomyrma* was possibly a ground forager, while *Haidomyrmex* and *Haidomyrmodes*, related genera in subfamily Sphecomyrminae, are reconstructed as active arboreal predators. Older ants in the genus *Sphecomyrmodes* have been found in 99 million year-old amber from Myanmar. A 2006 study suggested that ants arose tens of millions of years earlier than previously thought, up to 168 million years ago. After the rise of flowering plants about 100 million years ago they diversified and assumed ecological dominance around 60 million years ago. Some groups, such as the Leptanillinae and Martialinae, are suggested to have diversified from early primitive ants that were likely to have been predators underneath the surface of the soil.

Ants fossilised in Baltic amber.

During the Cretaceous period, a few species of primitive ants ranged widely on the Laurasian supercontinent (the Northern Hemisphere). They were scarce in comparison to the populations of other insects, representing only about 1% of the entire insect population. Ants became dominant after adaptive radiation at the beginning of the Paleogene period. By the Oligocene and Miocene, ants had come to represent 20–40% of all insects found in major fossil deposits. Of the species that lived in the Eocene epoch, around one in 10 genera survive to the present. Genera surviving today comprise 56% of the genera in Baltic amber fossils (early Oligocene), and 92% of the genera in Dominican amber fossils (apparently early Miocene).

Termites live in colonies and are sometimes called 'white ants', but termites are not ants. They are the sub-order Isoptera, and together with cockroaches they form the order Blattodea. Blattodeans are related to mantids, crickets, and other winged insects that do not undergo full metamorphosis. Like ants, termites are eusocial, with sterile workers, but they differ greatly in the genetics of reproduction. The similarity of their social structure to that of ants is attributed to convergent evolution. Velvet ants look like large ants, but are wingless female wasps.

Distribution and Diversity

Region	Number of species
Neotropics	2,162
Nearctic	580
Europe	180
Africa	2,500
Asia	2,080
Melanesia	275
Australia	985
Polynesia	42

Ants are found on all continents except Antarctica, and only a few large islands, such as Greenland, Iceland, parts of Polynesia and the Hawaiian Islands lack native ant species. Ants occupy a wide range of ecological niches and exploit many different food resources as direct or indirect herbivores, predators and scavengers. Most ant species are omnivorous generalists, but a few are specialist feeders. Their ecological dominance is demonstrated by their biomass: ants are estimated to contribute 15–20 % (on average and nearly 25% in the tropics) of terrestrial animal biomass, exceeding that of the vertebrates.

Ants range in size from 0.75 to 52 millimetres (0.030–2.0 in), the largest species being the fossil *Titanomyrma giganteum*, the queen of which was 6 centimetres (2.4 in) long with a wingspan of 15 centimetres (5.9 in). Ants vary in colour; most ants are red or black, but a few species are green and some tropical species have a metallic lustre. More than 12,000 species are currently known (with upper estimates of the potential existence of about 22,000) (see the article List of ant genera), with the greatest diversity in the tropics. Taxonomic studies continue to resolve the classification and systematics of ants. Online databases of ant species, including AntBase and the Hymenoptera Name Server, help to keep track of the known and newly described species. The relative ease with which ants may be sampled and studied in ecosystems has made them useful as indicator species in biodiversity studies.

Morphology

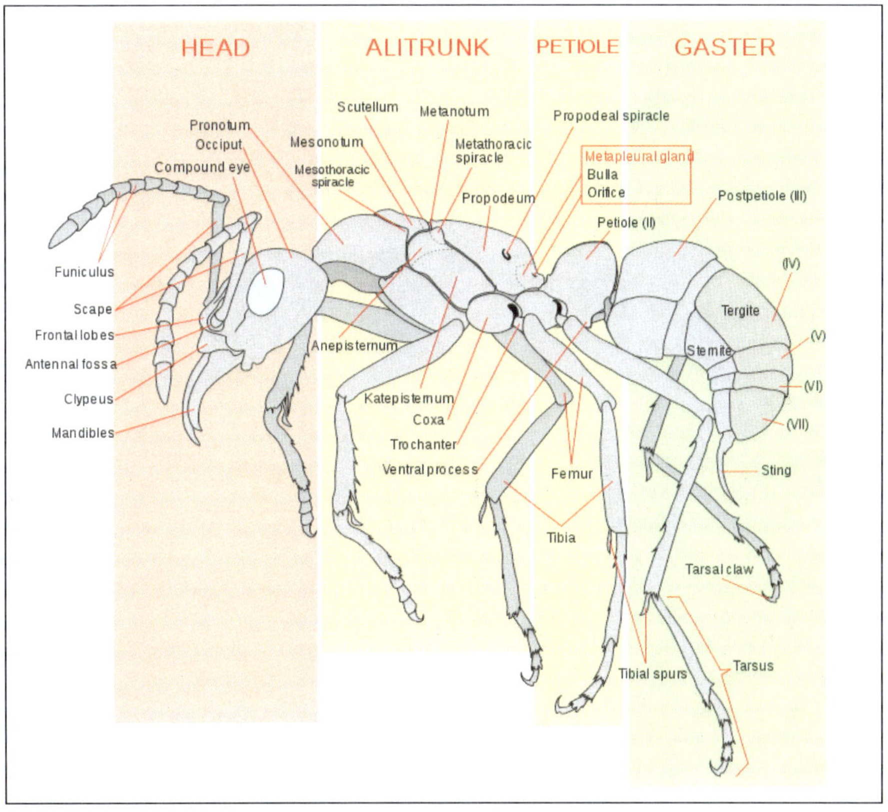

Diagram of a worker ant (*Pachycondyla verenae*).

Ants are distinct in their morphology from other insects in having elbowed antennae, metapleural glands, and a strong constriction of their second abdominal segment into a node-like petiole. The head, mesosoma, and metasoma are the three distinct body segments (formally tagmata). The petiole forms a narrow waist between their mesosoma (thorax plus the first abdominal segment, which is fused to it) and gaster (abdomen less the abdominal segments in the petiole). The petiole may be formed by one or two nodes (the second alone, or the second and third abdominal segments).

Like other insects, ants have an exoskeleton, an external covering that provides a protective casing around the body and a point of attachment for muscles, in contrast to the internal skeletons of humans and other vertebrates. Insects do not have lungs; oxygen and other gases, such as carbon dioxide, pass through their exoskeleton via tiny valves called spiracles. Insects also lack closed blood vessels; instead, they have a long, thin, perforated tube along the top of the body (called the "dorsal aorta") that functions like a heart, and pumps haemolymph toward the head, thus driving the circulation of the internal fluids. The nervous system consists of a ventral nerve cord that runs the length of the body, with several ganglia and branches along the way reaching into the extremities of the appendages.

Head

Macro portrait of an ant.

Bull ant showing the powerful mandibles and the relatively large compound eyes that provide excellent vision.

An ant's head contains many sensory organs. Like most insects, ants have compound eyes made from numerous tiny lenses attached together. Ant eyes are good for acute movement detection, but do not offer a high resolution image. They also have three small ocelli (simple eyes) on the top of the head that detect light levels and polarization. Compared to vertebrates, most ants have poor-to-mediocre eyesight and a few subterranean species are completely blind. However, some ants, such as Australia's bulldog ant, have excellent vision and are capable of discriminating the distance and size of objects moving nearly a metre away.

Two antennae ("feelers") are attached to the head; these organs detect chemicals, air currents, and vibrations; they also are used to transmit and receive signals through touch. The head has two strong jaws, the mandibles, used to carry food, manipulate objects, construct nests, and for defence. In some species, a small pocket (infrabuccal chamber) inside the mouth stores food, so it may be passed to other ants or their larvae.

Mesosoma

Both the legs and wings of the ant are attached to the mesosoma ("thorax"). The legs terminate in a hooked claw which allows them to hook on and climb surfaces. Only reproductive ants, queens, and males, have wings. Queens shed their wings after the nuptial flight, leaving visible stubs, a distinguishing feature of queens. In a few species, wingless queens (ergatoids) and males occur.

Metasoma

The metasoma (the "abdomen") of the ant houses important internal organs, including those of the reproductive, respiratory (tracheae), and excretory systems. Workers of many species have their egg-laying structures modified into stings that are used for subduing prey and defending their nests.

Polymorphism

Seven leafcutter ant workers of various castes (left) and two queens (right).

In the colonies of a few ant species, there are physical castes—workers in distinct size-classes, called minor, median, and major ergates. Often, the larger ants have disproportionately larger heads, and correspondingly stronger mandibles. These are known as macrergates while smaller workers are known as micrergates. Although formally known as dinergates, such individuals are sometimes called "soldier" ants because their stronger mandibles make them more effective in fighting, although they still are workers and their "duties" typically do not vary greatly from the minor or median workers. In a few species, the median workers are absent, creating a sharp divide between the minors and majors. Weaver ants, for example, have a distinct bimodal size distribution. Some other species show continuous variation in the size of workers. The smallest and largest workers in *Pheidologeton diversus* show nearly a 500-fold difference in their dry-weights.

Workers cannot mate; however, because of the haplodiploid sex-determination system in ants, workers of a number of species can lay unfertilised eggs that become fully fertile, haploid males. The role of workers may change with their age and in some species, such as honeypot ants, young workers are fed until their gasters are distended, and act as living food storage vessels. These food storage workers are called *repletes*. For instance, these replete workers develop in the North American honeypot ant *Myrmecocystus mexicanus*. Usually the largest workers in the colony develop into repletes; and, if repletes are removed from the colony, other workers become repletes, demonstrating the flexibility of this particular polymorphism. This polymorphism in morphology and behavior of workers initially was thought to be determined by environmental factors such as nutrition and hormones that led to different developmental paths; however, genetic differences between worker castes have been noted in *Acromyrmex* sp. These polymorphisms are caused by relatively small genetic changes; differences in a single gene of *Solenopsis invicta* can decide whether the colony will have single or multiple queens. The Australian jack jumper ant (*Myrmecia pilosula*) has only a single pair of chromosomes (with the males having just one chromosome as they are haploid), the lowest number known for any animal, making it an interesting subject for studies in the genetics and developmental biology of social insects.

Life Cycle

Meat eater ant nest during swarming.

The life of an ant starts from an egg. If the egg is fertilised, the progeny will be female diploid; if not, it will be male haploid. Ants develop by complete metamorphosis with the larva stages passing through a pupal stage before emerging as an adult. The larva is largely immobile and is fed and cared for by workers. Food is given to the larvae by trophallaxis, a process in which an ant regurgitates liquid food held in its crop. This is also how adults share food, stored in the "social stomach". Larvae, especially in the later stages, may also be provided solid food, such as trophic eggs, pieces of prey, and seeds brought by workers.

The larvae grow through a series of four or five moults and enter the pupal stage. The pupa has the appendages free and not fused to the body as in a butterfly pupa. The differentiation into queens and workers (which are both female), and different castes of workers, is influenced in some species by the nutrition the larvae obtain. Genetic influences and the control of gene expression by the developmental environment are complex and the determination of caste continues to be a subject of research. Winged male ants, called drones, emerge from pupae along with the usually winged breeding females. Some species, such as army ants, have wingless queens. Larvae and pupae need to be kept at fairly constant temperatures to ensure proper development, and so often, are moved around among the various brood chambers within the colony.

A new ergate spends the first few days of its adult life caring for the queen and young. She then graduates to digging and other nest work, and later to defending the nest and foraging. These changes are sometimes fairly sudden, and define what are called temporal castes. An explanation for the sequence is suggested by the high casualties involved in foraging, making it an acceptable risk only for ants who are older and are likely to die soon of natural causes.

Ant colonies can be long-lived. The queens can live for up to 30 years, and workers live from 1 to 3 years. Males, however, are more transitory, being quite short-lived and surviving for only a few weeks. Ant queens are estimated to live 100 times as long as solitary insects of a similar size.

Ants are active all year long in the tropics, but, in cooler regions, they survive the winter in a state of dormancy known as hibernation. The forms of inactivity are varied and some temperate species have larvae going into the inactive state (diapause), while in others, the adults alone pass the winter in a state of reduced activity.

Alate male ant, *Prenolepis imparis*.

Reproduction

Ants mating.

A wide range of reproductive strategies have been noted in ant species. Females of many species are known to be capable of reproducing asexually through thelytokous parthenogenesis. Secretions from the male accessory glands in some species can plug the female genital opening and prevent females from re-mating. Most ant species have a system in which only the queen and breeding females have the ability to mate. Contrary to popular belief, some ant nests have multiple queens, while others may exist without queens. Workers with the ability to reproduce are called "gamergates" and colonies that lack queens are then called gamergate colonies; colonies with queens are said to be queen-right.

Drones can also mate with existing queens by entering a foreign colony. When the drone is initially attacked by the workers, it releases a mating pheromone. If recognized as a mate, it will be carried to the queen to mate. Males may also patrol the nest and fight others by grabbing them with their mandibles, piercing their exoskeleton and then marking them with a pheromone. The marked male is interpreted as an invader by worker ants and is killed.

Fertilised meat-eater ant queen beginning to dig a new colony.

Most ants are univoltine, producing a new generation each year. During the species-specific breeding period, winged females and winged males, known to entomologists as alates, leave the colony in what is called a nuptial flight. The nuptial flight usually takes place in the late spring or early summer when the weather is hot and humid. Heat makes flying easier and freshly fallen rain makes the ground softer for mated queens to dig nests. Males typically take flight before the females. Males then use visual cues to find a common mating ground, for example, a landmark such as a pine tree to which other males in the area converge. Males secrete a mating pheromone that females follow. Males will mount females in the air, but the actual mating process usually takes place on the ground. Females of some species mate with just one male but in others they may mate with as many as ten or more different males, storing the sperm in their spermathecae.

Mated females then seek a suitable place to begin a colony. There, they break off their wings and begin to lay and care for eggs. The females can selectively fertilise future eggs with the sperm stored to produce diploid workers or lay unfertilized haploid eggs to produce drones. The first workers to hatch are known as nanitics, and are weaker and smaller than later workers, but they begin to serve the colony immediately. They enlarge the nest, forage for food, and care for the other eggs. Species that have multiple queens may have a queen leaving the nest along with some workers to found a colony at a new site, a process akin to swarming in honeybees.

Behavior and Ecology

Communication

Two *Camponotus sericeus* workers communicating through touch and pheromones.

Ants communicate with each other using pheromones, sounds, and touch. The use of pheromones as chemical signals is more developed in ants, such as the red harvester ant, than in other hymenopteran groups. Like other insects, ants perceive smells with their long, thin, and mobile antennae. The paired antennae provide information about the direction and intensity of scents. Since most ants live on the ground, they use the soil surface to leave pheromone trails that may be followed by other ants. In species that forage in groups, a forager that finds food marks a trail on the way back to the colony; this trail is followed by other ants, these ants then reinforce the trail when they head back with food to the colony. When the food source is exhausted, no new trails are marked by returning ants and the scent slowly dissipates. This behavior helps ants deal with changes in their environment. For instance, when an established path to a food source is blocked by an obstacle, the foragers leave the path to explore new routes. If an ant is successful, it leaves a new trail marking the shortest route on its return. Successful trails are followed by more ants, reinforcing better routes and gradually identifying the best path.

Ants use pheromones for more than just making trails. A crushed ant emits an alarm pheromone that sends nearby ants into an attack frenzy and attracts more ants from farther away. Several ant species even use "propaganda pheromones" to confuse enemy ants and make them fight among themselves. Pheromones are produced by a wide range of structures including Dufour's glands, poison glands and glands on the hindgut, pygidium, rectum, sternum, and hind tibia. Pheromones also are exchanged, mixed with food, and passed by trophallaxis, transferring information within the colony. This allows other ants to detect what task group (e.g., foraging or nest maintenance) other colony members belong to. In ant species with queen castes, when the dominant queen stops producing a specific pheromone, workers begin to raise new queens in the colony.

Some ants produce sounds by stridulation, using the gaster segments and their mandibles. Sounds may be used to communicate with colony members or with other species.

Defence

A *Plectroctena* sp. attacks another of its kind to protect its territory.

Ants attack and defend themselves by biting and, in many species, by stinging, often injecting or spraying chemicals, such as formic acid in the case of formicine ants, alkaloids and piperidines in fire ants, and a variety of protein components in other ants. Bullet ants (*Paraponera*), located in Central and South America, are considered to have the most painful sting of any insect, although it is usually not fatal to humans. This sting is given the highest rating on the Schmidt Sting Pain Index.

The sting of jack jumper ants can be fatal, and an antivenom has been developed for it.

Fire ants, *Solenopsis* spp., are unique in having a venom sac containing piperidine alkaloids. Their stings are painful and can be dangerous to hypersensitive people.

A weaver ant in fighting position, mandibles wide open.

Trap-jaw ants of the genus *Odontomachus* are equipped with mandibles called trap-jaws, which snap shut faster than any other predatory appendages within the animal kingdom. One study of *Odontomachus bauri* recorded peak speeds of between 126 and 230 km/h (78 and 143 mph), with the jaws closing within 130 microseconds on average. The ants were also observed to use their jaws as a catapult to eject intruders or fling themselves backward to escape a threat. Before striking, the ant opens its mandibles extremely widely and locks them in this position by an internal mechanism. Energy is stored in a thick band of muscle and explosively released when triggered by the stimulation of sensory organs resembling hairs on the inside of the mandibles. The mandibles also permit slow and fine movements for other tasks. Trap-jaws also are seen in the following genera: *Anochetus*, *Orectognathus*, and *Strumigenys*, plus some members of the Dacetini tribe, which are viewed as examples of convergent evolution.

A Malaysian species of ant in the *Camponotus cylindricus* group has enlarged mandibular glands that extend into their gaster. If combat takes a turn for the worse, a worker may perform a final act of suicidal altruism by rupturing the membrane of its gaster, causing the content of its mandibular glands to burst from the anterior region of its head, spraying a poisonous, corrosive secretion containing acetophenones and other chemicals that immobilise small insect attackers. The worker subsequently dies.

Suicidal defences by workers are also noted in a Brazilian ant, *Forelius pusillus*, where a small group of ants leaves the security of the nest after sealing the entrance from the outside each evening.

Ant mound holes prevent water from entering the nest during rain.

In addition to defence against predators, ants need to protect their colonies from pathogens. Some worker ants maintain the hygiene of the colony and their activities include undertaking or *necrophory*, the disposal of dead nest-mates. Oleic acid has been identified as the compound released from dead ants that triggers necrophoric behavior in *Atta mexicana* while workers of *Linepithema humile* react to the absence of characteristic chemicals (dolichodial and iridomyrmecin) present on the cuticle of their living nestmates to trigger similar behavior.

Nests may be protected from physical threats such as flooding and overheating by elaborate nest architecture. Workers of *Cataulacus muticus*, an arboreal species that lives in plant hollows, respond to flooding by drinking water inside the nest, and excreting it outside. *Camponotus anderseni*, which nests in the cavities of wood in mangrove habitats, deals with submergence under water by switching to anaerobic respiration.

Learning

Many animals can learn behaviours by imitation, but ants may be the only group apart from mammals where interactive teaching has been observed. A knowledgeable forager of *Temnothorax albipennis* will lead a naive nest-mate to newly discovered food by the process of tandem running. The follower obtains knowledge through its leading tutor. The leader is acutely sensitive to the progress of the follower and slows down when the follower lags and speeds up when the follower gets too close.

Controlled experiments with colonies of *Cerapachys biroi* suggest that an individual may choose nest roles based on her previous experience. An entire generation of identical workers was divided into two groups whose outcome in food foraging was controlled. One group was continually rewarded with prey, while it was made certain that the other failed. As a result, members of the successful group intensified their foraging attempts while the unsuccessful group ventured out fewer and fewer times. A month later, the successful foragers continued in their role while the others had moved to specialise in brood care.

Nest Construction

Leaf nest of weaver ants, Pamalican, Philippines.

Complex nests are built by many ant species, but other species are nomadic and do not build permanent structures. Ants may form subterranean nests or build them on trees. These nests may be found in the ground, under stones or logs, inside logs, hollow stems, or even acorns. The materials used for construction include soil and plant matter, and ants carefully select their nest sites; *Temnothorax albipennis* will avoid sites with dead ants, as these may indicate the presence of pests or disease. They are quick to abandon established nests at the first sign of threats.

The army ants of South America, such as the *Eciton burchellii* species, and the driver ants of Africa do not build permanent nests, but instead, alternate between nomadism and stages where the workers form a temporary nest (bivouac) from their own bodies, by holding each other together.

Weaver ant (*Oecophylla* spp.) workers build nests in trees by attaching leaves together, first pulling them together with bridges of workers and then inducing their larvae to produce silk as they are moved along the leaf edges. Similar forms of nest construction are seen in some species of *Polyrhachis*.

Formica polyctena, among other ant species, constructs nests that maintain a relatively constant interior temperature that aids in the development of larvae. The ants maintain the nest temperature by choosing the location, nest materials, controlling ventilation and maintaining the heat from solar radiation, worker activity and metabolism, and in some moist nests, microbial activity in the nest materials.

Some ant species, such as those that use natural cavities, can be opportunistic and make use of the controlled micro-climate provided inside human dwellings and other artificial structures to house their colonies and nest structures.

Ant bridge.

Cultivation of Food

Most ants are generalist predators, scavengers, and indirect herbivores, but a few have evolved specialised ways of obtaining nutrition. It is believed that many ant species that engage in indirect herbivory rely on specialized symbiosis with their gut microbes to upgrade the nutritional value of the food they collect and allow them to survive in nitrogen poor regions, such as rainforest canopies. Leafcutter ants (*Atta* and *Acromyrmex*) feed exclusively on a fungus that grows only within their colonies. They continually collect leaves which are taken to the colony, cut into tiny pieces and placed in fungal gardens. Ergates specialise in related tasks according to their sizes. The largest ants cut stalks, smaller workers chew the leaves and the smallest tend the fungus. Leafcutter ants are sensitive enough to recognise the reaction of the fungus to different plant material, apparently detecting chemical signals from the fungus. If a particular type of leaf is found to be toxic to the fungus, the colony will no longer collect it. The ants feed on structures produced by the fungi called *gongylidia*. Symbiotic bacteria on the exterior surface of the ants produce antibiotics that kill bacteria introduced into the nest that may harm the fungi.

Myrmecocystus, honeypot ants, store food to prevent colony famine.

Navigation

An ant trail.

Foraging ants travel distances of up to 200 metres (700 ft) from their nest and scent trails allow them to find their way back even in the dark. In hot and arid regions, day-foraging ants face death by desiccation, so the ability to find the shortest route back to the nest reduces that risk. Diurnal desert ants of the genus *Cataglyphis* such as the Sahara desert ant navigate by keeping track of direction as well as distance travelled. Distances travelled are measured using an internal pedometer that keeps count of the steps taken and also by evaluating the movement of objects in their visual field (optical flow). Directions are measured using the position of the sun. They integrate this information to find the shortest route back to their nest. Like all ants, they can also make use of visual landmarks when available as well as olfactory and tactile cues to navigate. Some species of ant are able to use the Earth's magnetic field for navigation. The compound eyes of ants have specialised cells that detect polarised light from the Sun, which is used to determine direction. These polarization detectors are sensitive in the ultraviolet region of the light spectrum. In some army ant species, a group of foragers who become separated from the main column may sometimes turn back on themselves and form a circular ant mill. The workers may then run around continuously until they die of exhaustion.

Locomotion

The female worker ants do not have wings and reproductive females lose their wings after their mating flights in order to begin their colonies. Therefore, unlike their wasp ancestors, most ants travel by walking. Some species are capable of leaping. For example, Jerdon's jumping ant (*Harpegnathos saltator*) is able to jump by synchronising the action of its mid and hind pairs of legs. There are several species of gliding ant including *Cephalotes atratus*; this may be a common trait among arboreal ants with small colonies. Ants with this ability are able to control their horizontal movement so as to catch tree trunks when they fall from atop the forest canopy.

Other species of ants can form chains to bridge gaps over water, underground, or through spaces in vegetation. Some species also form floating rafts that help them survive floods. These rafts may also have a role in allowing ants to colonise islands. *Polyrhachis sokolova*, a species of ant found in Australian mangrove swamps, can swim and live in underwater nests. Since they lack gills, they go to trapped pockets of air in the submerged nests to breathe.

Cooperation and Competition

Meat-eater ants feeding on a cicada: social ants cooperate and collectively gather food.

Not all ants have the same kind of societies. The Australian bulldog ants are among the biggest and most basal of ants. Like virtually all ants, they are eusocial, but their social behavior is poorly developed compared to other species. Each individual hunts alone, using her large eyes instead of chemical senses to find prey.

Some species (such as *Tetramorium caespitum*) attack and take over neighbouring ant colonies. Others are less expansionist, but just as aggressive; they invade colonies to steal eggs or larvae, which they either eat or raise as workers or slaves. Extreme specialists among these slave-raiding ants, such as the Amazon ants, are incapable of feeding themselves and need captured workers to survive. Captured workers of enslaved *Temnothorax* species have evolved a counter strategy, destroying just the female pupae of the slave-making *Temnothorax americanus*, but sparing the males (who don't take part in slave-raiding as adults).

Ants identify kin and nestmates through their scent, which comes from hydrocarbon-laced secretions that coat their exoskeletons. If an ant is separated from its original colony, it will eventually lose the colony scent. Any ant that enters a colony without a matching scent will be attacked. Also, the reason why two separate colonies of ants will attack each other even if they are of the same species is because the genes responsible for pheromone production are different between them. The Argentine ant, however, does not have this characteristic, due to lack of genetic diversity, and has become a global pest because of it.

A worker *Harpegnathos saltator* (a jumping ant) engaged in battle with a rival colony's queen.

Parasitic ant species enter the colonies of host ants and establish themselves as social parasites; species such as *Strumigenys xenos* are entirely parasitic and do not have workers, but instead, rely on the food gathered by their *Strumigenys perplexa* hosts. This form of parasitism is seen across many ant genera, but the parasitic ant is usually a species that is closely related to its host. A variety of methods are employed to enter the nest of the host ant. A parasitic queen may enter the host nest before the first brood has hatched, establishing herself prior to development of a colony scent. Other species use pheromones to confuse the host ants or to trick them into carrying the parasitic queen into the nest. Some simply fight their way into the nest.

A conflict between the sexes of a species is seen in some species of ants with these reproducers apparently competing to produce offspring that are as closely related to them as possible. The most extreme form involves the production of clonal offspring. An extreme of sexual conflict is seen in *Wasmannia auropunctata*, where the queens produce diploid daughters by thelytokous parthenogenesis and males produce clones by a process whereby a diploid egg loses its maternal contribution to produce haploid males who are clones of the father.

Relationships with other Organisms

The spider *Myrmarachne plataleoides* (female shown) mimics weaver ants to avoid predators.

Ants form symbiotic associations with a range of species, including other ant species, other insects, plants, and fungi. They also are preyed on by many animals and even certain fungi. Some arthropod species spend part of their lives within ant nests, either preying on ants, their larvae, and eggs, consuming the food stores of the ants, or avoiding predators. These inquilines may bear a close resemblance to ants. The nature of this ant mimicry (myrmecomorphy) varies, with some cases involving Batesian mimicry, where the mimic reduces the risk of predation. Others show Wasmannian mimicry, a form of mimicry seen only in inquilines.

An ant collects honeydew from an aphid.

Aphids and other hemipteran insects secrete a sweet liquid called honeydew, when they feed on plant sap. The sugars in honeydew are a high-energy food source, which many ant species collect. In some cases, the aphids secrete the honeydew in response to ants tapping them with their antennae. The ants in turn keep predators away from the aphids and will move them from one feeding location to another. When migrating to a new area, many colonies will take the aphids with them, to ensure a continued supply of honeydew. Ants also tend mealybugs to harvest their honeydew. Mealybugs may become a serious pest of pineapples if ants are present to protect mealybugs from their natural enemies.

Myrmecophilous (ant-loving) caterpillars of the butterfly family Lycaenidae (e.g., blues, coppers, or hairstreaks) are herded by the ants, led to feeding areas in the daytime, and brought inside the ants' nest at night. The caterpillars have a gland which secretes honeydew when the ants massage them. Some caterpillars produce vibrations and sounds that are perceived by the ants. A similar adaptation can be seen in Grizzled skipper butterflies that emit vibrations by expanding their wings in order to communicate with ants, which are natural predators of these butterflies. Other caterpillars have evolved from ant-loving to ant-eating: these myrmecophagous caterpillars secrete a pheromone that makes the ants act as if the caterpillar is one of their own larvae. The caterpillar is then taken into the ant nest where it feeds on the ant larvae. A number of specialized bacterial have been found as endosymbionts in ant guts. Some of the dominant bacteria belong to the order Rhizobiales whose members are known for being nitrogen-fixing symbionts in legumes but the species found in ant lack the ability to fix nitrogen. Fungus-growing ants that make up the tribe Attini, including leafcutter ants, cultivate certain species of fungus in the genera *Leucoagaricus* or *Leucocoprinus* of the family Agaricaceae. In this ant-fungus mutualism, both species depend on each other for survival. The ant *Allomerus decemarticulatus* has evolved a three-way association with the host plant, *Hirtella physophora* (Chrysobalanaceae), and a sticky fungus which is used to trap their insect prey.

Ants may obtain nectar from flowers such as the dandelion but are only rarely known to pollinate flowers.

Lemon ants make devil's gardens by killing surrounding plants with their stings and leaving a pure patch of lemon ant trees, (*Duroia hirsuta*). This modification of the forest provides the ants with more nesting sites inside the stems of the *Duroia* trees. Although some ants obtain nectar from flowers, pollination by ants is somewhat rare, one example being of the pollination of the orchid *Leporella fimbriata* which induces male *Myrmecia urens* to pseudocopulate with the flowers, transferring pollen in the process. One theory that has been proposed for the rarity of pollination is that the secretions of the metapleural gland inactivate and reduce the viability of pollen. Some plants have special nectar exuding structures, extrafloral nectaries, that provide food for ants, which in turn protect the plant from more damaging herbivorous insects. Species such as the bullhorn acacia (*Acacia cornigera*) in Central America have hollow thorns that house colonies of stinging ants (*Pseudomyrmex ferruginea*) who defend the tree against insects, browsing mammals, and epiphytic vines. Isotopic labelling studies suggest that plants also obtain nitrogen from the ants. In return, the ants obtain food from protein- and lipid-rich Beltian bodies. In Fiji *Philidris nagasau* (Dolichoderinae) are known to selectively grow species of epiphytic *Squamellaria* (Rubiaceae) which produce large domatia inside which the ant colonies nest. The ants plant the seeds and the domatia of young seedling are immediately occupied and the ant faeces in them contribute to rapid growth. Similar dispersal associations are found with other dolichoderines in the region as well. Another example of this type of ectosymbiosis comes from the *Macaranga* tree, which has stems adapted to house colonies of *Crematogaster* ants.

Many plant species have seeds that are adapted for dispersal by ants. Seed dispersal by ants or myrmecochory is widespread, and new estimates suggest that nearly 9% of all plant species may have such ant associations. Often, seed-dispersing ants perform directed dispersal, depositing the seeds in locations that increase the likelihood of seed survival to reproduction. Some plants in arid, fire-prone systems are particularly dependent on ants for their survival and dispersal as the seeds are transported to safety below the ground. Many ant-dispersed seeds have special external structures, elaiosomes, that are sought after by ants as food.

A convergence, possibly a form of mimicry, is seen in the eggs of stick insects. They have an edible elaiosome-like structure and are taken into the ant nest where the young hatch.

A meat ant tending a common leafhopper nymph.

Most ants are predatory and some prey on and obtain food from other social insects including other ants. Some species specialise in preying on termites (*Megaponera* and *Termitopone*) while a few Cerapachyinae prey on other ants. Some termites, including *Nasutitermes corniger*, form associations with certain ant species to keep away predatory ant species. The tropical wasp *Mischocyttarus drewseni* coats the pedicel of its nest with an ant-repellent chemical. It is suggested that many tropical wasps may build their nests in trees and cover them to protect themselves from ants. Other wasps, such as *A. multipicta*, defend against ants by blasting them off the nest with bursts of wing buzzing. Stingless bees (*Trigona* and *Melipona*) use chemical defences against ants.

Flies in the Old World genus *Bengalia* (Calliphoridae) prey on ants and are kleptoparasites, snatching prey or brood from the mandibles of adult ants. Wingless and legless females of the Malaysian phorid fly (*Vestigipoda myrmolarvoidea*) live in the nests of ants of the genus *Aenictus* and are cared for by the ants.

Fungi in the genera *Cordyceps* and *Ophiocordyceps* infect ants. Ants react to their infection by climbing up plants and sinking their mandibles into plant tissue. The fungus kills the ants, grows on their remains, and produces a fruiting body. It appears that the fungus alters the behavior of the ant to help disperse its spores in a microhabitat that best suits the fungus. Strepsipteran parasites also manipulate their ant host to climb grass stems, to help the parasite find mates.

A nematode (*Myrmeconema neotropicum*) that infects canopy ants (*Cephalotes atratus*) causes the black-coloured gasters of workers to turn red. The parasite also alters the behavior of the ant, causing them to carry their gasters high. The conspicuous red gasters are mistaken by birds for ripe fruits, such as *Hyeronima alchorneoides*, and eaten. The droppings of the bird are collected by other ants and fed to their young, leading to further spread of the nematode.

Spiders sometimes feed on ants.

South American poison dart frogs in the genus *Dendrobates* feed mainly on ants, and the toxins in their skin may come from the ants.

Army ants forage in a wide roving column, attacking any animals in that path that are unable to escape. In Central and South America, *Eciton burchellii* is the swarming ant most commonly attended by "ant-following" birds such as antbirds and woodcreepers. This behavior was once considered mutualistic, but later studies found the birds to be parasitic. Direct kleptoparasitism (birds stealing food from the ants' grasp) is rare and has been noted in Inca doves which pick seeds at nest entrances as they are being transported by species of *Pogonomyrmex*. Birds that follow ants eat many prey insects and thus decrease the foraging success of ants. Birds indulge in a peculiar behavior called anting that, as yet, is not fully understood. Here birds rest on ant nests, or pick and drop ants onto their wings and feathers; this may be a means to remove ectoparasites from the birds.

Anteaters, aardvarks, pangolins, echidnas and numbats have special adaptations for living on a diet of ants. These adaptations include long, sticky tongues to capture ants and strong claws to break into ant nests. Brown bears (*Ursus arctos*) have been found to feed on ants. About 12%, 16%, and 4% of their faecal volume in spring, summer, and autumn, respectively, is composed of ants.

Carcinology

Carcinology is a branch of zoology that consists of the study of crustaceans, a group of arthropods that includes lobsters, crayfish, shrimp, krill, barnacles and crabs. Other names for carcinology are malacostracology, crustaceology, and crustalogy, and a person who studies crustaceans is a *carcinologist* or occasionally a *malacostracologist*, a *crustaceologist*, or a *crustalogist*.

Crayfish

Crayfish are freshwater crustaceans resembling small lobsters (to which they are related). They are also known as crawfish, crawdads, freshwater lobsters, mountain lobsters, mudbugs, or yabbies. Taxonomically, they are members of the superfamilies Astacoidea and Parastacoidea. They breathe through feather-like gills. Some species are found in brooks and streams where fresh water is running, while others thrive in swamps, ditches, and paddy fields. Most crayfish cannot tolerate polluted water, although some species such as *Procambarus clarkii* are hardier. Crayfish feed on animals and plants, either living or decomposing, and detritus.

Names

The name "crayfish" comes from the Old French word *escrevisse* (Modern French *écrevisse*). The word has been modified to "crayfish" by association with "fish" (folk etymology). The largely American variant "crawfish" is similarly derived.

Some kinds of crayfish are known locally as lobsters, crawdads, mudbugs, and yabbies. In the Eastern United States, "crayfish" is more common in the north, while "crawdad" is heard more in central and southwestern regions, and "crawfish" further south, although considerable overlaps exist.

The study of crayfish is called astacology.

Other Animals

In Australia (on the eastern seaboard), New Zealand and South Africa, the term *crayfish* or *cray* generally refers to a saltwater spiny lobster, of the genus *Jasus* that is indigenous to much of southern Oceania, while the freshwater species are usually called *yabbies* or *kōura*, from the indigenous Australian and Māori names for the animal respectively, or by other names specific to each species. Exceptions include western rock lobster (of the Palinuridae family) found on the west coast of Australia; the Tasmanian giant freshwater crayfish (from the Parastacidae family) found only in Tasmania; and the Murray crayfish found along Australia's Murray River.

In Singapore, the term *crayfish* typically refers to *Thenus orientalis*, a seawater crustacean from the slipper lobster family. True crayfish are not native to Singapore, but are commonly found as pets, or as an invasive species (*Cherax quadricarinatus*) in the many water catchment areas, and are alternatively known as *freshwater lobsters*.

Anatomy

The body of a decapod crustacean, such as a crab, lobster, or prawn (shrimp), is made up of twenty body segments grouped into two main body parts, the cephalothorax and the abdomen. Each segment may possess one pair of appendages, although in various groups these may be reduced or missing. On average, crayfish grow to 17.5 centimetres (6.9 in) in length, but some grow larger. Walking legs have a small claw at the end.

Geographical Distribution and Classification

The Three Families of Crayfish

Astacidae: *Austropotamobius pallipes.*

Cambaridae: *Procambarus alleni.*

Parastacidae: *Cherax pulcher.*

There are three families of crayfish, two in the Northern Hemisphere and one in the Southern Hemisphere. The Southern Hemisphere (Gondwana-distributed) family Parastacidae, with 14 extant genera and two extinct genera, live(d) in South America, Madagascar and Australasia. They are distinguished by the absence of the first pair of pleopods. Of the other two families, the three genera of the Astacidae live in western Eurasia and western North America, while the 15 genera of the family Cambaridae live in eastern Asia and eastern North America.

North America

The greatest diversity of crayfish species is found in southeastern North America, with over 330 species in nine genera, all in the family Cambaridae. A further genus of astacid crayfish is found in the Pacific Northwest and the headwaters of some rivers east of the Continental Divide. Many crayfish are also found in lowland areas where the water is abundant in calcium, and oxygen rises from underground springs.

In 1983, Louisiana designated the crayfish, or crawfish as they are commonly called, as its official state crustacean. Louisiana produces 100 million pounds of crawfish per year with the red swamp and white river crawfish being the main species harvested. Crawfish are a part of Cajun culture dating back hundreds of years. A variety of cottage industries have developed as a result of commercialized crawfish iconology. Their products include crawfish attached to wooden plaques, T-shirts with crawfish logos, and crawfish pendants, earrings and necklaces made of gold or silver.

Australia

Australia has over 100 species in a dozen genera. Australia is home to the world's three largest freshwater crayfish – the Tasmanian giant freshwater crayfish *Astacopsis gouldi*, which can achieve a mass of over 5 kilograms (11 lb) and is found in rivers of northern Tasmania, the Murray crayfish *Euastacus armatus*, which can reach 2.5 kilograms (5.5 lb), (although there have been reports of animals up to 3 kilograms (6.6 lb)) and is found in much of the southern Murray-Darling basin and the marron from Western Australia (now believed to be two species, *Cherax tenuimanus* and *C. cainii*) which may reach a weight of 2.2 kilograms (4.9 lb).

Many of the better-known Australian crayfish are of the genus *Cherax*, and include the common yabby (*Cherax destructor*), western yabby (*Cherax preissii*) and red-claw crayfish (*Cherax quadricarinatus*).

The marron species *C. tenuimanus* is critically endangered, while other large Australasian crayfish are threatened or endangered.

New Zealand

In New Zealand, two species of *Paranephrops* are endemic, and are known by the Māori name *kōura*.

Fossil Record

Fossil records of crayfish older than 30 million years are rare, but fossilised burrows have been found from strata as old as the late Palaeozoic or early Mesozoic. The oldest records of the Parastacidae are in Australia, and are 115 million years old.

Crayfish Plague

Some crayfish suffer from a disease called crayfish plague, caused by the North American water mould *Aphanomyces astaci* which was transmitted to Europe when North American species of crayfish were introduced there. Species of the genus *Astacus* are particularly susceptible to infection, allowing the plague-coevolved signal crayfish to invade parts of Europe.

Uses

Human Uses

Crayfish, boiled with potatoes and corn.

A pet crayfish, *Procambarus clarkii* in a freshwater aquarium.

Golden crayfish pendant, Chiriqui, Panama.

Food

Crayfish are eaten worldwide. Like other edible crustaceans, only a small portion of the body of a crayfish is eaten. In most prepared dishes, such as soups, bisques and étouffées, only the tail portion is served. At crawfish boils or other meals where the entire body of the crayfish is presented, other portions, such as the claw meat, may be eaten. Like all crustaceans, crayfish are not kosher because they are aquatic animals that do not have both fins and scales. They are therefore not eaten by observant Jews.

As of 2005, Louisiana supplies 95% of the crayfish harvested in the US. In 1987, Louisiana produced 90% of the crayfish harvested in the world, 70% of which were consumed locally. In 2007, the Louisiana crawfish harvest was about 54,800 tons, almost all of it from aquaculture. About 70%–80% of crayfish produced in Louisiana are *Procambarus clarkii* (red swamp crawfish), with the remaining 20%–30% being *Procambarus zonangulus* (white river crawfish).

Bait

Crayfish are preyed upon by a variety of ray-finned fishes, and are commonly used as bait, either live or with only the tail meat. They are a popular bait for catching catfish, largemouth bass, smallmouth bass, striped bass, perch, pike and muskie. When using live crayfish as bait, anglers prefer to hook them between the eyes, piercing through their hard, pointed beak which causes them no harm; therefore, they remain more active.

When using crayfish as bait, it is important to fish in the same environment where they were caught. An Illinois State University report that focused on studies conducted on the Fox River and Des Plaines River watershed stated that rusty crayfish, initially caught as bait in a different environment, were dumped into the water and "outcompeted the native clearwater crayfish." Other studies confirmed that transporting crayfish to different environments have led to various ecological problems, including the elimination of native species. Transporting crayfishes as live bait has also contributed to the spread of zebra mussels in various waterways throughout Europe and North America, as they are known to attach themselves to exoskeleton of crayfishes.

Pets

Crayfish are kept as pets in freshwater aquariums. They prefer foods like shrimp pellets or various vegetables, but will also eat tropical fish food, regular fish food, algae wafers, and small fish that can be captured with their claws. A report by the National Park Service as well as video and anecdotal reports by aquarium owners indicate that crayfish will eat their molted exoskeleton "to recover the calcium and phosphates contained in it." As omnivores, crayfish will eat almost anything; therefore, they may explore the edibility of aquarium plants in a fish tank. However, most species of dwarf crayfish, such as *Cambarellus patzcuarensis*, will not destructively dig or eat live aquarium plants.

In some nations, such as the United Kingdom, United States, Australia, and New Zealand, imported alien crayfish are a danger to local rivers. The three species commonly imported to Europe from the Americas are *Orconectes limosus*, *Pacifastacus leniusculus* and *Procambarus clarkii*. Crayfish may spread into different bodies of water because specimens captured for pets in one river are often released into a different catchment. There is a potential for ecological damage when crayfish are introduced into non-native bodies of water: e.g., crayfish plague in Europe, or the introduction of the common yabby (*Cherax destructor*) into drainages east of the Great Dividing Range in Australia.

Sentinel Species

The Protivin brewery in the Czech Republic uses crayfish outfitted with sensors to detect any changes in their bodies or pulse activity in order to monitor the purity of the water used in their product. The creatures are kept in a fish tank that is fed with the same local natural source water used in their brewing. If three or more of the crayfish have changes to their pulses, employees know there is a change in the water and examine the parameters.

Scientists also monitor crayfish in the wild in natural bodies of water to study the levels of pollutants there.

Barnacle

A barnacle is a type of arthropod constituting the infraclass Cirripedia in the subphylum Crustacea, and is hence related to crabs and lobsters. Barnacles are exclusively marine, and tend to live in shallow and tidal waters, typically in erosive settings. They are sessile (nonmobile) suspension feeders, and have four nektonic (active swimming) larval stages. Around 70 barnacle species are currently known. The name "Cirripedia" is Greek, meaning "curl-footed". The study of barnacles is called cirripedology.

Whale barnacles attached to the throat of a humpback whale.

Barnacles are encrusters, attaching themselves temporarily to a hard substrate. The most common, "acorn barnacles" (Sessilia), are sessile, growing their shells directly onto the substrate. The order *Pedunculata* (goose barnacles and others) attach themselves by means of a stalk.

Free-living barnacles are attached to the substratum by cement glands that form the base of the first pair of antennae; in effect, the animal is fixed upside down by means of its forehead. In some barnacles, the cement glands are fixed to a long, muscular stalk, but in most they are part of a flat membrane or calcified plate. A ring of plates surrounds the body, homologous with the carapace of other crustaceans. These consist of the rostrum, two lateral plates, two carinolaterals, and a carina. In sessile barnacles, the apex of the ring of plates is covered by an operculum, which may be recessed into the carapace. The plates are held together by various means, depending on species, in some cases being solidly fused.

Inside the carapace, the animal lies on its stomach, with its limbs projecting downwards. Segmentation is usually indistinct, and the body is more or less evenly divided between the head and thorax, with little, if any, abdomen. Adult barnacles have few appendages on their heads, with only a single, vestigial pair of antennae, attached to the cement gland. The eight pairs of thoracic limbs are referred to as "cirre", which are feathery and very long, being used to filter food, such as plankton, from the water and move it towards the mouth.

Barnacles have no true heart, although a sinus close to the esophagus performs a similar function, with blood being pumped through it by a series of muscles. The blood vascular system is minimal. Similarly, they have no gills, absorbing oxygen from the water through their limbs and the inner membrane of their carapaces. The excretory organs of barnacles are maxillary glands.

The main sense of barnacles appears to be touch, with the hairs on the limbs being especially sensitive. The adult also has three photoreceptors (ocelli), one median and two lateral. These photoreceptors record the stimulus for the barnacle shadow reflex, where a sudden decrease in light causes cessation of the fishing rhythm and closing of the opercular plates. The photoreceptors are likely only capable of sensing the difference between light and dark. This eye is derived from the primary naupliar eye.

Life Cycle

Barnacles have two distinct larval stages, the nauplius and the cyprid, before developing into a mature adult.

Nauplius

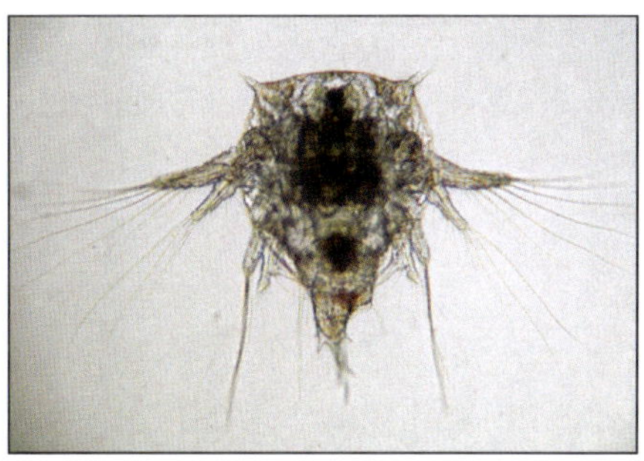

Nauplius larva of *Elminius modestus*.

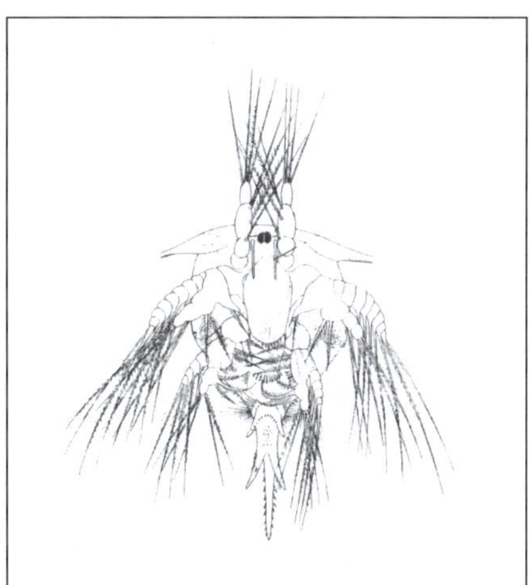

Nauplius larva of a barnacle with fronto-lateral horns

A fertilised egg hatches into a nauplius: a one-eyed larva comprising a head and a telson, without a thorax or abdomen. This undergoes six moults, passing through five instars, before transforming into the cyprid stage. Nauplii are typically initially brooded by the parent, and released after the first moult as larvae that swim freely using setae.

Cyprid

The cyprid larva is the last larval stage before adulthood. It is not a feeding stage; its role is to find a suitable place to settle, since the adults are sessile. The cyprid stage lasts from days to weeks. It explores potential surfaces with modified antennules; once it has found a potentially suitable spot, it attaches head-first using its antennules and a secreted glycoproteinous substance. Larvae assess surfaces based upon their surface texture, chemistry, relative wettability, color, and the presence or absence and composition of a surface biofilm; swarming species are also more likely to attach near other barnacles. As the larva exhausts its finite energy reserves, it becomes less selective in the sites it selects. It cements itself permanently to the substrate with another proteinaceous compound, and then undergoes metamorphosis into a juvenile barnacle.

Adult

Typical acorn barnacles develop six hard calcareous plates to surround and protect their bodies. For the rest of their lives, they are cemented to the substrate, using their feathery legs (cirri) to capture plankton.

Once metamorphosis is over and they have reached their adult form, barnacles continue to grow by adding new material to their heavily calcified plates. These plates are not moulted; however, like all ecdysozoans, the barnacle itself will still moult its cuticle.

Sexual Reproduction

Most barnacles are hermaphroditic, although a few species are gonochoric or androdioecious. The ovaries are located in the base or stalk, and may extend into the mantle, while the testes are towards the back of the head, often extending into the thorax. Typically, recently moulted hermaphroditic individuals are receptive as females. Self-fertilization, although theoretically possible, has been experimentally shown to be rare in barnacles.

The sessile lifestyle of barnacles makes sexual reproduction difficult, as the organisms cannot leave their shells to mate. To facilitate genetic transfer between isolated individuals, barnacles have extraordinarily long penises. Barnacles probably have the largest penis to body size ratio of the animal kingdom.

Barnacles can also reproduce through a method called spermcasting, in which the male barnacle releases his sperm into the water and females pick it up and fertilise their eggs.

The Rhizocephala superorder used to be considered hermaphroditic, but it turned out that its males inject themselves into the female's body, degrading to the condition of nothing more than sperm-producing cells.

Ecology

Most barnacles are suspension feeders; they dwell continually in their shells, which are usually constructed of six plates, and reach into the water column with modified legs. These feathery appendages beat rhythmically to draw plankton and detritus into the shell for consumption.

Other members of the class have quite a different mode of life. For example, members of the super-order Rhizocephala, including the genus *Sacculina*, are parasitic and live within crabs.

Although they have been found at water depths to 600 m (2,000 ft), most barnacles inhabit shallow waters, with 75% of species living in water depths less than 100 m (300 ft), and 25% inhabiting the intertidal zone. Within the intertidal zone, different species of barnacles live in very tightly constrained locations, allowing the exact height of an assemblage above or below sea level to be precisely determined.

Since the intertidal zone periodically desiccates, barnacles are well adapted against water loss. Their calcite shells are impermeable, and they possess two plates which they can slide across their apertures when not feeding. These plates also protect against predation.

Barnacles are displaced by limpets and mussels, which compete for space. They also have numerous predators. They employ two strategies to overwhelm their competitors: "swamping" and fast growth. In the swamping strategy, vast numbers of barnacles settle in the same place at once, covering a large patch of substrate, allowing at least some to survive in the balance of probabilities. Fast growth allows the suspension feeders to access higher levels of the water column than their competitors, and to be large enough to resist displacement; species employing this response, such as the aptly named *Megabalanus*, can reach 7 cm (3 in) in length; other species may grow larger still (*Austromegabalanus psittacus*).

Competitors may include other barnacles, and disputed evidence indicates balanoid barnacles competitively displaced chthalamoid barnacles. Balanoids gained their advantage over the chthalamoids in the Oligocene, when they evolved tubular skeletons, which provide better anchorage to the substrate, and allow them to grow faster, undercutting, crushing, and smothering chthalamoids.

Among the most common predators on barnacles are whelks. They are able to grind through the calcareous exoskeletons of barnacles and feed on the softer inside parts. Mussels also prey on barnacle larvae. Another predator on barnacles is the starfish species *Pisaster ochraceus*.

Barnacles and limpets compete for space in the intertidal zone.

Goose barnacles, with their cirri extended for feeding.

Underside of large *Chesaconcavus* sp. (Miocene) showing internal plates in bioimmured smaller barnacles.

The anatomy of parasitic barnacles is generally simpler than that of their free-living relatives. They have no carapace or limbs, having only unsegmented sac-like bodies. Such barnacles feed by extending thread-like rhizomes of living cells into their hosts' bodies from their points of attachment.

Classification

Some authorities regard the Cirripedia as a full class or subclass, and the orders listed above are sometimes treated as superorders. In 2001, Martin and Davis placed Cirripedia as an infraclass of Thecostraca and divided it into six orders:

Infraclass Cirripedia Burmeister, 1834

- Superorder Acrothoracica Gruvel, 1905
  - Order Pygophora Berndt, 1907
  - Order Apygophora Berndt, 1907
- Superorder Rhizocephala Müller, 1862
  - Order Kentrogonida Delage, 1884
  - Order Akentrogonida Häfele, 1911
- Superorder Thoracica Darwin, 1854
  - Order Pedunculata Lamarck, 1818
  - Order Sessilia Lamarck, 1818

Fossil Record

The geological history of barnacles can be traced back to animals such as *Priscansermarinus* from the Middle Cambrian (on the order of 510 to 500 million years ago), although they do not become common as skeletal remains in the fossil record until the Neogene (last 20 million years). In part, their poor skeletal preservation is due to their restriction to high-energy environments, which tend to be erosional – therefore it is more common for their shells to be ground up by wave action than for them to reach a depositional setting. Trace fossils of acrothoracican barnacle borings (*Rogerella*) are common in the fossil record from the Devonian to the recent.

Barnacles can play an important role in estimating paleo-water depths. The degree of disarticulation of fossils suggests the distance they have been transported, and since many species have narrow ranges of water depths, it can be assumed that the animals lived in shallow water and broke up as they were washed down-slope. The completeness of fossils, and nature of damage, can thus be used to constrain the tectonic history of regions.

Balanus improvisus, one of the many barnacle taxa described by Charles Darwin.

Miocene (Messinian) *Megabalanus*, smothered by sand and fossilised.

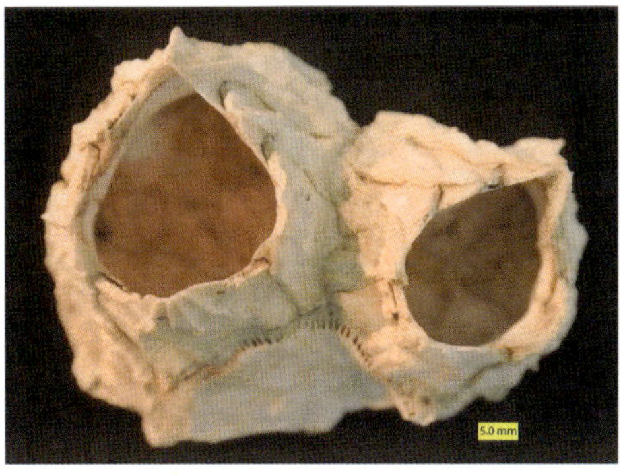

Chesaconcavus, a Miocene barnacle from Maryland.

Relationship with Humans

Barnacles are of economic consequence, as they often attach themselves to synthetic structures, sometimes to the structure's detriment. Particularly in the case of ships, they are classified as fouling organisms. The amount and size of barnacles that cover ships can negatively impact their efficiency.

The stable isotope signals in the layers of barnacle shells can potentially be used as a forensic tracking method for whales, loggerhead turtles and marine debris, such as shipwrecks or a flaperon suspected to be from Malaysia Airlines Flight 370.

Some barnacles are considered edible by humans, including Japanese goose barnacles (*e.g. Capitulum mitella*), and goose barnacles (*e.g. Pollicipes pollicipes*), a delicacy in Spain and Portugal. The resemblance of this barnacle's fleshy stalk to a goose's neck gave rise, in ancient times, to the notion that geese literally grew from the barnacle. Indeed, the word "barnacle" originally referred to a species of goose, the barnacle goose *Branta leucopsis*, whose eggs and young were rarely seen by humans because it breeds in the remote Arctic.

Additionally, the picoroco barnacle is used in Chilean cuisine and is one of the ingredients in *curanto*.

Copepod

Copepods are a group of small crustaceans found in nearly every freshwater and saltwater habitat. Some species are planktonic (drifting in sea waters), some are benthic (living on the ocean floor), and some continental species may live in limnoterrestrial habitats and other wet terrestrial places, such as swamps, under leaf fall in wet forests, bogs, springs, ephemeral ponds, and puddles, damp moss, or water-filled recesses (phytotelmata) of plants such as bromeliads and pitcher plants. Many live underground in marine and freshwater caves, sinkholes, or stream beds. Copepods are sometimes used as biodiversity indicators.

As with other crustaceans, copepods have a larval form. For copepods, the egg hatches into a nauplius form, with a head and a tail but no true thorax or abdomen. The larva molts several times until it resembles the adult and then, after more molts, achieves adult development. The nauplius form is so different from the adult form that it was once thought to be a separate species.

Classification and Diversity

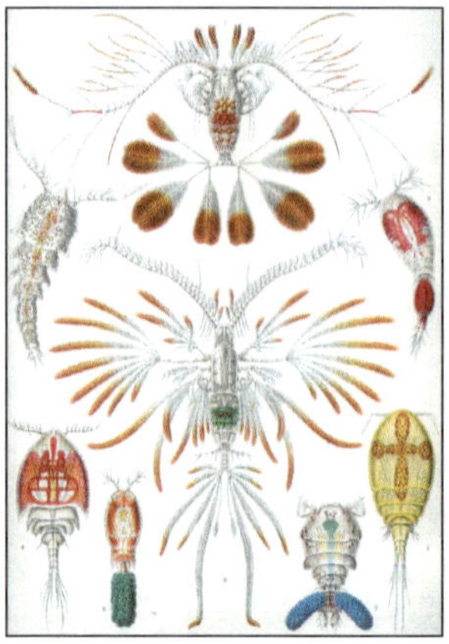

Copepods from Ernst Haeckel's *Kunstformen der Natur*.

Copepods form a subclass belonging to class Hexanauplia in the subphylum Crustacea (crustaceans); they are divided into 10 orders. Some 13,000 species of copepods are known, and 2,800 of them live in fresh water.

Characteristics

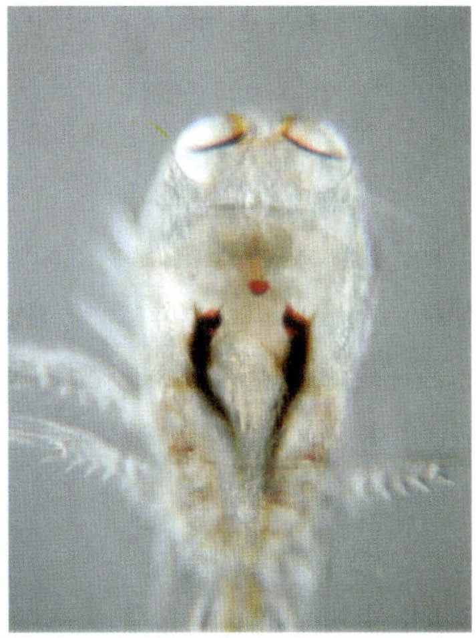

Most copepods have a single naupliar eye in the middle of their head, but copepods of the genera *Copilia* and *Corycaeus* possess two eyes. Each eye has a large anterior cuticular lens paired with a posterior internal lens to form a telescope.

Copepods vary considerably, but can typically be 1 to 2 mm (0.04 to 0.08 in) long, with a teardrop-shaped body and large antennae. Like other crustaceans, they have an armoured exoskeleton, but they are so small that in most species, this thin armour and the entire body is almost totally transparent. Some polar copepods reach 1 cm (0.39 in). Most copepods have a single median compound eye, usually bright red and in the centre of the transparent head; subterranean species may be eyeless. Like other crustaceans, copepods possess two pairs of antennae; the first pair is often long and conspicuous.

Free-living copepods of the orders Calanoida, Cyclopoida, and Harpacticoida typically have a short, cylindrical body, with a rounded or beaked head, although considerable variation exists in this pattern. The head is fused with the first one or two thoracic segments, while the remainder of the thorax has three to five segments, each with limbs. The first pair of thoracic appendages is modified to form maxillipeds, which assist in feeding. The abdomen is typically narrower than the thorax, and contains five segments without any appendages, except for some tail-like "rami" at the tip. Parasitic copepods (the other seven orders) vary widely in morphology and no generalizations are possible.

Because of their small size, copepods have no need of any heart or circulatory system (the members of the order Calanoida have a heart, but no blood vessels), and most also lack gills. Instead, they absorb oxygen directly into their bodies. Their excretory system consists of maxillary glands.

Behavior

The second pair of cephalic appendages in free-living copepods is usually the main time-averaged source of propulsion, beating like oars to pull the animal through the water. However, different groups have different modes of feeding and locomotion, ranging from almost immotile for several minutes (e.g. some harpacticoid copepods) to intermittent motion (e.g., some cyclopoid copepods) and continuous displacements with some escape reactions (e.g. most calanoid copepods.)

Some copepods have extremely fast escape responses when a predator is sensed, and can jump with high speed over a few millimetres. Many species have neurons surrounded by myelin (for increased conduction speed), which is very rare among invertebrates (other examples are some annelids and malacostracan crustaceans like palaemonid shrimp and penaeids). Even rarer, the myelin is highly organized, resembling the well-organized wrapping found in vertebrates (Gnathostomata). Despite their fast escape response, copepods are successfully hunted by slow-swimming seahorses, which approach their prey so gradually, it senses no turbulence, then suck the copepod into their snout too suddenly for the copepod to escape.

Finding a mate in the three-dimensional space of open water is challenging. Some copepod females solve the problem by emitting pheromones, which leave a trail in the water that the male can follow. Copepods experience a low Reynolds number and therefore a high relative viscosity. One foraging strategy involves chemical detection of sinking marine snow aggregates and taking advantage of nearby low-pressure gradients to swim quickly towards food sources.

Diet

Most free-living copepods feed directly on phytoplankton, catching cells individually. A single copepod can consume up to 373,000 phytoplanktons per day. They generally have to clear the equivalent to about a million times their own body volume of water every day to cover their nutritional needs. Some of the larger species are predators of their smaller relatives. Many benthic copepods eat organic detritus or the bacteria that grow in it, and their mouth parts are adapted for scraping and biting. Herbivorous copepods, particularly those in rich, cold seas, store up energy from their food as oil droplets while they feed in the spring and summer on plankton blooms. These droplets may take up over half of the volume of their bodies in polar species. Many copepods (e.g., fish lice like the Siphonostomatoida) are parasites, and feed on their host organisms. In fact, three of the 10 known orders of copepods are wholly or largely parasitic, with another three comprising most of the free-living species.

Life Cycle

Egg sac of a copepod.

Most nonparasitic copepods are holoplanktonic, meaning they stay planktonic for all of their lifecycles, although harpacticoids, although free-living, tend to be benthic rather than planktonic. During mating, the male copepod grips the female with his first pair of antennae, which is sometimes modified for this purpose. The male then produces an adhesive package of sperm and transfers it to the female's genital opening with his thoracic limbs. Eggs are sometimes laid directly into the water, but many species enclose them within a sac attached to the female's body until they hatch. In some pond-dwelling species, the eggs have a tough shell and can lie dormant for extended periods if the pond dries up.

Eggs hatch into nauplius larvae, which consist of a head with a small tail, but no thorax or true abdomen. The nauplius moults five or six times, before emerging as a "copepodid larva". This stage resembles the adult, but has a simple, unsegmented abdomen and only three pairs of thoracic limbs. After a further five moults, the copepod takes on the adult form. The entire process from hatching to adulthood can take a week to a year, depending on the species and environmental conditions such as temperature and nutrition (e.g., egg-to-adult time in the calanoid *Parvocalanus crassirostris* is ~7 days at 25 °C but 19 days at 15 °C.

Ecology

Lernaeolophus sultanus (Pennellidae), parasite of the fish *Pristipomoides filamentosus*, scale: each division = 1 mm.

Planktonic copepods are important to global ecology and the carbon cycle. They are usually the dominant members of the zooplankton, and are major food organisms for small fish such as the dragonet, banded killifish, Alaska pollock, and other crustaceans such as krill in the ocean and in fresh water. Some scientists say they form the largest animal biomass on earth. Copepods compete for this title with Antarctic krill (*Euphausia superba*). *C. glacialis* inhabits the edge of the Arctic icepack, especially in polynyas where light (and photosynthesis) is present, in which they alone comprise up to 80% of zooplankton biomass. They bloom as the ice recedes each spring. The ongoing large reductions in the annual minimum of recent years may force them to compete in the open ocean with the much less nourishing *C. finmarchicus*, which is spreading from the North Sea and the Norwegian Sea into the Barents Sea.

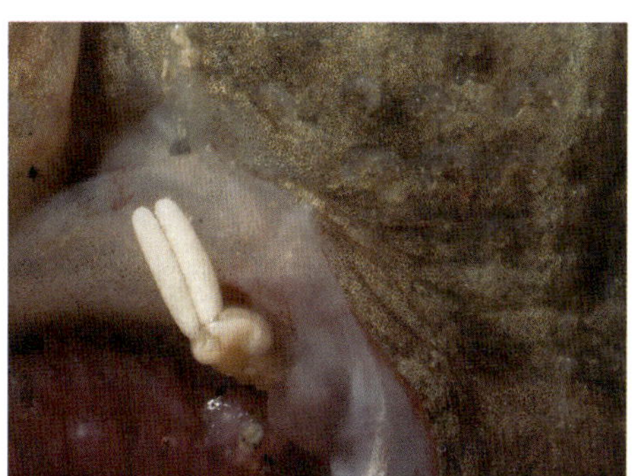

Acanthochondria cornuta, an ectoparasite on flounder in the North Sea.

Because of their smaller size and relatively faster growth rates, and because they are more evenly distributed throughout more of the world's oceans, copepods almost certainly contribute far more to the secondary productivity of the world's oceans, and to the global ocean carbon sink than krill, and perhaps more than all other groups of organisms together. The surface layers of the oceans are currently believed to be the world's largest carbon sink, absorbing about 2 billion tons of carbon a year, the equivalent to perhaps a third of human carbon emissions, thus reducing their impact. Many planktonic copepods feed near the surface at night, then sink (by changing oils into more dense fats) into deeper water during the day to avoid visual predators. Their moulted exoskeletons, faecal pellets, and respiration at depth all bring carbon to the deep sea.

About half of the estimated 13,000 described species of copepods are parasitic and have strongly modified bodies. They attach themselves to bony fish, sharks, marine mammals, and many kinds of invertebrates such as molluscs, tunicates, or corals. They live as endo- or ectoparasites on fish or invertebrates in fresh water and in marine environments.

Copepods as Parasitic Hosts

In addition to being parasites themselves, copepods are subject to parasitic infection. The most common parasite is the marine dinoflagellates, *Blastodinium* spp., which are gut parasites of many copepod species. Currently, 12 species of *Blastodinium* are described, the majority of which were discovered in the Mediterranean Sea. Most *Blastodinium* species infect several different hosts, but species-specific infection of copepods does occur. Generally, adult copepod females and juveniles are infected.

During the naupliar stage, the copepod host ingests the unicellular dinospore of the parasite. The dinospore is not digested and continues to grow inside the intestinal lumen of the copepod. Eventually, the parasite divides into a multicellular arrangement called a trophont. This trophont is considered parasitic, contains thousands of cells, and can be several hundred micrometers in length. The trophont is greenish to brownish in color as a result of well-defined chloroplasts. At maturity, the trophont ruptures and *Blastodinium* spp. are released from the copepod anus as free dinospore cells. Not much is known about the dinospore stage of *Blastodinium* and its ability to persist outside of the copepod host in relatively high abundances.

The copepod *Calanus finmarchicus*, which dominates the northeastern Atlantic coast, has been shown to be greatly infected by this parasite. A 2014 study in this region found up to 58% of collected *C. finmarchicus* females to be infected. In this study, *Blastodinium*-infected females had no measurable feeding rate over a 24-hour period. This is compared to uninfected females which, on average, ate 2.93×10^4 cells copepod^{-1} d^{-1}. *Blastodinium*-infected females of *C. finmarchicus* exhibited characteristic signs of starvation, including decreased respiration, fecundity, and fecal pellet production. Though photosynthetic, *Blastodinium* spp. procure most of their energy from organic material in the copepod gut, thus contributing to host starvation. Underdeveloped or disintegrated ovaries, as well as decreased fecal pellet size, are a direct result of starvation in female copepods. Infection from *Blastodinium* spp. could have serious ramifications on the success of copepod species and the function of entire marine ecosystems. Parasitism via *Blastodinium* spp.' is not lethal, but has negative impacts on copepod physiology, which in turn may alter marine biogeochemical cycles.

Freshwater copepods of the *Cyclops* genus are the intermediate host of *Dracunculus medinensis*, the Guinea worm nematode that causes dracunculiasis disease in humans. This disease may be close to being eradicated through efforts at the U.S. Centers for Disease Control and Prevention and the World Health Organization.

Practical Aspects

Marine Aquaria

Live copepods are used in the saltwater aquarium hobby as a food source and are generally considered beneficial in most reef tanks. They are scavengers and also may feed on algae, including coralline algae. Live copepods are popular among hobbyists who are attempting to keep particularly difficult species such as the mandarin dragonet or scooter blenny. They are also popular to hobbyists who want to breed marine species in captivity. In a saltwater aquarium, copepods are typically stocked in the refugium.

Water Supplies

Copepods are sometimes found in public main water supplies, especially systems where the water is not mechanically filtered, such as New York City, Boston, and San Francisco. This is not usually a problem in treated water supplies. In some tropical countries, such as Peru and Bangladesh, a correlation has been found between copepods' presence and cholera in untreated water, because the cholera bacteria attach to the surfaces of planktonic animals. The larvae of the guinea worm must develop within a copepod's digestive tract before being transmitted to humans. The risk of infection with these diseases can be reduced by filtering out the copepods (and other matter), for example with a cloth filter.

Copepods have been used successfully in Vietnam to control disease-bearing mosquitoes such as *Aedes aegypti* that transmit dengue fever and other human parasitic diseases.

The copepods can be added to water-storage containers where the mosquitoes breed. Copepods, primarily of the genera *Mesocyclops* and *Macrocyclops* (such as *Macrocyclops albidus*), can survive for periods of months in the containers, if the containers are not completely drained by their users. They attack, kill, and eat the younger first- and second-instar larvae of the mosquitoes. This biological control method is complemented by community trash removal and recycling to eliminate other possible mosquito-breeding sites. Because the water in these containers is drawn from uncontaminated sources such as rainfall, the risk of contamination by cholera bacteria is small, and in fact no cases of cholera have been linked to copepods introduced into water-storage containers. Trials using copepods to control container-breeding mosquitoes are underway in several other countries, including Thailand and the southern United States. The method, though, would be very ill-advised in areas where the guinea worm is endemic.

The presence of copepods in the New York City water supply system has caused problems for some Jewish people who observe kashrut. Copepods, being crustaceans, are not kosher, nor are they small enough to be ignored as nonfood microscopic organisms, since some specimens can be seen with the naked eye. When a group of rabbis in Brooklyn, New York, discovered the copepods in the summer of 2004, they triggered such enormous debate in rabbinic circles that some observant Jews felt compelled to buy and install filters for their water. The water was ruled kosher by posek Yisrael Belsky.

Helminthology

The helminth *Spinochordodes* parasitising a bush-cricket (*Meconema* sp.).

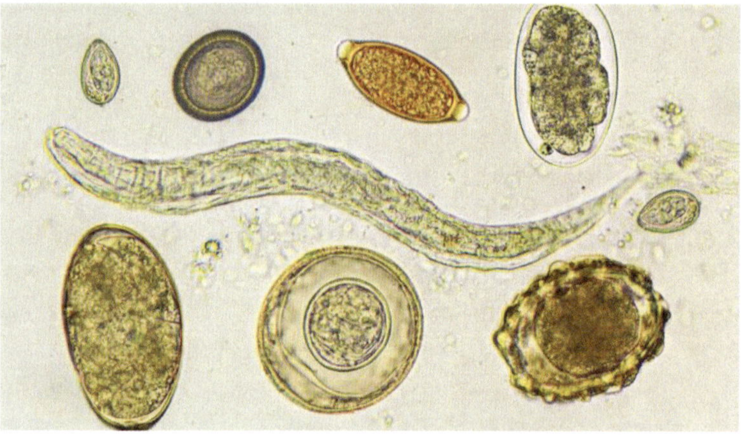

Helminthology is the study of parasitic worms (helminths). The field studies the taxonomy of helminths and their effects on their hosts.

In the 18th and early 19th century there was wave of publications on helminthology; this period has been described as the science's "Golden Era". During that period the authors Félix Dujardin, William Blaxland Benham, Peter Simon Pallas, Marcus Elieser Bloch, Otto Friedrich Müller, Johann Goeze, Friedrich Zenker, Charles Wardell Stiles, Carl Asmund Rudolphi, Otto Friedrich Bernhard von Linstow and Johann Gottfried Bremser started systematic scientific studies of the subject.

The Japanese parasitologist Satyu Yamaguti was one of the most active helminthologists of the 20th century; he wrote the six-volume *Systema Helminthum*.

Malacology

Malacology, is the study of Mollusca, such as this bigfin reef squid.

Malacology is the branch of invertebrate zoology that deals with the study of the Mollusca (mollusks or molluscs), the second-largest phylum of animals in terms of described species after the arthropods. Mollusks include snails and slugs, clams, octopus and squid, and numerous other kinds, many of which have shells. One division of malacology, conchology, is devoted to the study of mollusk shells.

Fields within malacological research include taxonomy, ecology and evolution. Applied malacology studies medical, veterinary, and agricultural applications, for example mollusks as vectors of disease, as in schistosomiasis.

Archaeology employs malacology to understand the evolution of the climate, the biota of the area, and the usage of the site.

Conchology

Calliostoma tigris.

Conchology is the study of mollusc shells. Conchology is one aspect of malacology, the study of molluscs; however, malacology is the study of molluscs as whole organisms, whereas conchology is confined to the study of their shells. It includes the study of land and freshwater mollusc shells as well as seashells and extends to the study of a gastropod's operculum.

Shell of *Lobatus gigas*, the queen conch.

Conchology is now sometimes seen as an archaic study, because relying on only one aspect of an organism's morphology can be misleading. However, a shell often gives at least some insight into molluscan taxonomy, and historically the shell was often the only part of exotic species that was available for study. Even in current museum collections it is common for the dry material (shells) to greatly exceed the amount of material that is preserved whole in alcohol.

Conchologists mainly deal with four molluscan orders: the gastropods (snails), bivalves (clams), Polyplacophora (chitons) and Scaphopoda (tusk shells). Cephalopods only have small internal shells, with the exception of the Nautiloidea. Some groups, such as the sea slug nudibranchs, have lost their shells altogether, while in others it has been replaced by a protein support structure.

Invertebrate Paleontology

Invertebrate Paleontology is the study of invertebrate animals of the geologic past. Invertebrates constitute about 95% of all living animal species; and, although they share the common negative characteristic of the lack of a backbone or vertebral column (cf. Chordata), they show very great variation in other respects. Living invertebrates include such widely different forms as the sponges, corals, clams, snails, squids, crabs, lobsters, insects, starfish, and sea urchins. They also live in a great variety of different environments, ranging from the ocean depths to mountainous areas and from the polar regions to equatorial jungles. They show great diversity in ways of locomotion and food gathering; some are parasitic. About three-quarters of the total of about 1,120,000 species of living invertebrates are insects.

References

- Arthropodology, dictionary: biology-online.org, Retrieved 02 June, 2019
- Barker, S. C.; Murrell, A. (2004). "Systematics and evolution of ticks with a list of valid genus and species names". Parasitology. 129 (S1): S15–S36
- Barth, Friedrich G. (2013). A Spider's World: Senses and Behavior. Springer Science & Business Media. ISBN 9783662048993
- Jason A. Dunlop; David Penney; O. Erik Tetlie & Lyall I. Anderson (2008). "How many species of fossil arachnids are there" (PDF). Journal of Arachnology. 36 (2): 262–272. doi:10.1636/CH07-89.1

3

Anatomy of Invertebrates

Anatomy is the branch of biology that deals with the study of the structures of the living organisms. Porifera, cnidaria, echinoderm, platyhelminthes, nematodes, annelida, arthropoda, mollusca, etc. are some of the invertebrates studied within this field. This chapter delves into anatomy of these invertebrates to provide an extensive understanding of this discipline.

Anatomy of Porifera

Unlike Protozoans, the Poriferans are multicellular. However, unlike higher metazoans, the cells that make up a sponge are not organized into tissues. Therefore, sponges lack true tissues and organs; in addition, they have no body symmetry. Sponges do, however, have specialized cells that perform specific functions. The shapes of their bodies are adapted for maximal efficiency of water flow through the central cavity, where nutrients are deposited, and leaves through a hole called the osculum. Many sponges have internal skeletons of spongin and/or spicules of calcium carbonate or silica. Primarily, their body consists of a thin sheet of cells over a frame (skeleton). As their name suggests, Poriferans are characterized by the presence of minute pores called ostia on their body.

Since water is vital to sponges for excretion, feeding, and gas exchange, their body structure facilitates the movement of water through the sponge. Structures such as canals, chambers, and cavities enable water to move through the sponge to nearly all body cells.

Most species use sexual reproduction, releasing sperm cells into the water to fertilize ova that in some species are released and in others are retained by the "mother." The fertilized eggs form larvae which swim off in search of places to settle. Sponges are also known for regenerating from fragments that are broken off, although this only works if the fragments include the right types of cells. A few species reproduce by budding. When conditions deteriorate, such as when temperatures drop, many freshwater species and a few marine ones produce gemmules: "survival pods" of unspecialized cells that remain dormant until conditions improve. They then either form completely new sponges or recolonize the skeletons of their parents.

Most of the approximately 5,000–10,000 known species of sponges are filter-feeders, feeding on bacteria and other food particles in the water. However, a few species of sponge that live in food-poor environments have become carnivores that prey mainly on small crustaceans. Other species host photosynthesizing micro-organisms as endosymbionts; these alliances often produce more food and oxygen than they consume.

Morphology of Sponges

Instead of true tissues or organs, sponges have specialized cells that are in charge of important bodily functions and processes.

The morphology of the simplest sponges takes the shape of a cylinder with a large central cavity, the spongocoel, occupying the inside of the cylinder. Water can enter into the spongocoel from numerous pores in the body wall. Water entering the spongocoel is extruded via a large, common opening called the osculum. However, sponges exhibit a range of diversity in body forms, including variations in the size of the spongocoel, the number of osculi, and where the cells that filter food from the water are located.

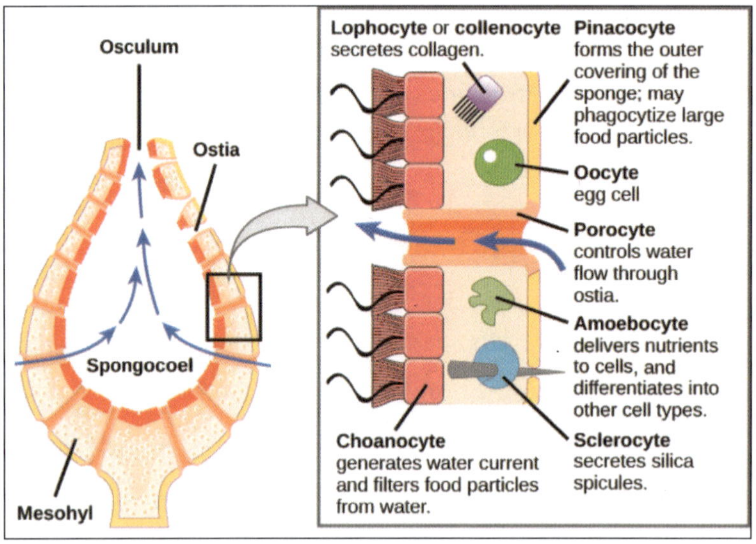

Sponge morphology: The sponge's (a) basic body plan is a cylinder shape with a large central cavity. The specialized cell types in sponges (b) each perform a distinct function.

While sponges (excluding the Hexactinellids) do not exhibit tissue-layer organization, they do have different cell types that perform distinct functions. Pinacocytes, which are epithelial-like cells, form the outermost layer of sponges, enclosing a jelly-like substance called mesohyl. Mesohyl is an extracellular matrix consisting of a collagen -like gel with suspended cells that perform various functions. The gel-like consistency of mesohyl acts as an endoskeleton, maintaining the tubular morphology of sponges. In addition to the osculum, sponges have multiple pores called ostia on their bodies that allow water to enter the sponge. In some sponges, ostia are formed by porocytes: single, tube-shaped cells that act as valves to regulate the flow of water into the spongocoel. In other sponges, ostia are formed by folds in the body wall of the sponge.

Choanocytes ("collar cells") are present at various locations, depending on the type of sponge; however, they always line the inner portions of some space through which water flows: the spongocoel in simple sponges; canals within the body wall in more complex sponges; and chambers scattered throughout the body in the most complex sponges. Whereas pinacocytes line the outside of the sponge, choanocytes tend to line certain inner portions of the sponge body that surround the mesohyl. The structure of a choanocyte is critical to its function, which is to generate a water current through the sponge and to trap and ingest food particles by phagocytosis. Note that there is a similarity in appearance between the sponge choanocyte and choanoflagellates (Protista). This similarity suggests that sponges and choanoflagellates are closely related and probably share a recent, common ancestry.

The cell body is embedded in mesohyl. It contains all organelles required for normal cell function, but protruding into the "open space" inside of the sponge is a mesh-like collar composed of microvilli with a single flagellum in the center of the column. The cumulative effect of the flagella from all choanocytes aids the movement of water through the sponge: drawing water into the sponge through the numerous ostia, into the spaces lined by choanocytes, and eventually out through the osculum (or osculi). Meanwhile, food particles, including waterborne bacteria and algae, are trapped by the sieve-like collar of the choanocytes, slide down into the body of the cell, are ingested by phagocytosis, and become encased in a food vacuole. Finally, choanocytes will differentiate into sperm for sexual reproduction; they will become dislodged from the mesohyl, leaving the sponge with expelled water through the osculum.

The second crucial cells in sponges are called amoebocytes (or archaeocytes), named for the fact that they move throughout the mesohyl in an amoeba-like fashion. Amoebocytes have a variety of functions: delivering nutrients from choanocytes to other cells within the sponge; giving rise to eggs for sexual reproduction (which remain in the mesohyl); delivering phagocytized sperm from choanocytes to eggs; and differentiating into more-specific cell types. Some of these more-specific cell types include collencytes and lophocytes, which produce the collagen-like protein to maintain the mesohyl; sclerocytes, which produce spicules in some sponges; and spongocytes, which produce the protein spongin in the majority of sponges. These cells produce collagen to maintain the consistency of the mesohyl.

Physiological Processes in Sponges

Sponges are sessile, feed by phagocytosis, and reproduce sexually and asexually; all major functions are regulated by water flow diffusion.

Sponges, despite being simple organisms, regulate their different physiological processes through a variety of mechanisms. These mechanisms regulate metabolism, reproduction, and locomotion.

Metabolism

Sponges lack complex digestive, respiratory, circulatory, reproductive, and nervous systems. Their food is trapped when water passes through the ostia and out through the osculum. Bacteria smaller than 0.5 microns in size are trapped by choanocytes, which are the principal cells engaged in nutrition, and are ingested by phagocytosis. Particles that are larger than the ostia may be phagocytized by pinacocytes. In some sponges, amoebocytes transport food from cells that have ingested food particles to those that do not. For this type of digestion, in which food particles are digested within individual cells, the sponge draws water through diffusion. The limit of this type of digestion is that food particles must be smaller than individual cells.

All other major body functions in the sponge (gas exchange, circulation, excretion) are performed by diffusion between the cells that line the openings within the sponge and the water that is passing through those openings. All cell types within the sponge obtain oxygen from water through diffusion. Likewise, carbon dioxide is released into seawater by diffusion. In addition, nitrogenous waste produced as a by-product of protein metabolism is excreted via diffusion by individual cells into the water as it passes through the sponge.

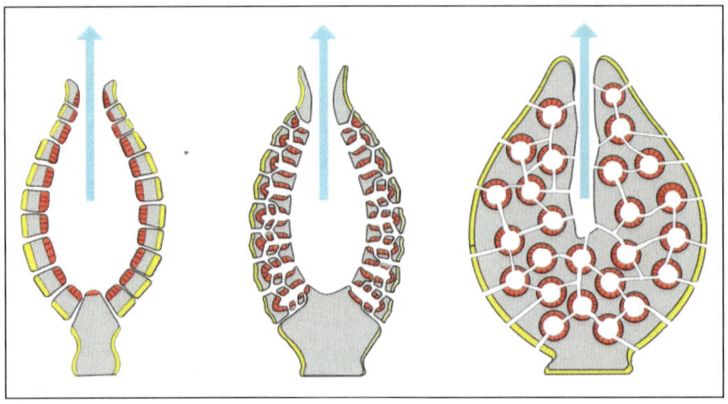

Water flow in a sponge: In a sponge, water enters through the body pores and exits in the direction of the osculum (direction of blue arrow). This diffusion of water through the body supports major functions in the sponge.

Locomotion

Sponges are generally sessile as adults and spend their lives attached to a fixed substratum. They do not show movement over large distances as do free-swimming marine invertebrates. However, sponge cells are capable of creeping along substrata via organizational plasticity. Under experimental conditions, researchers have shown that sponge cells spread on a physical support demonstrate a leading edge for directed movement. It has been speculated that this localized creeping movement may help sponges adjust to microenvironments near the point of attachment. It must be noted, however, that this pattern of movement has been documented in laboratories, but it remains to be observed in natural sponge habitats.

Anatomy of Cnidaria

Basic Body Forms

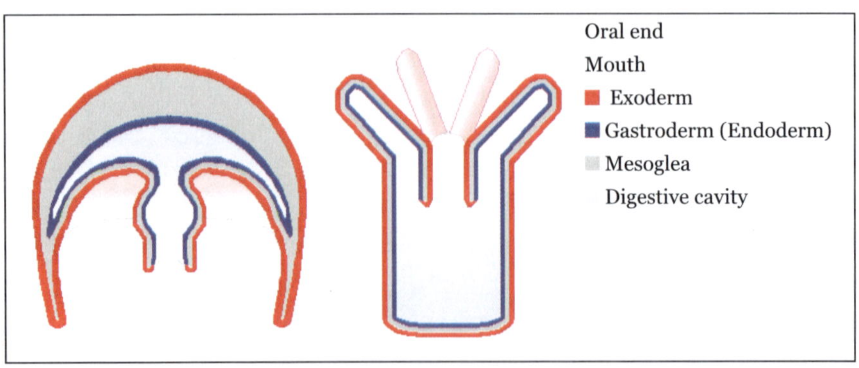

Aboral end Oral end.

Most adult cnidarians appear as either free-swimming medusae or sessile polyps, and many hydrozoans species are known to alternate between the two forms.

Both are radially symmetrical, like a wheel and a tube respectively. Since these animals have no heads, their ends are described as "oral" (nearest the mouth) and "aboral" (furthest from the mouth).

Most have fringes of tentacles equipped with cnidocytes around their edges, and medusae generally have an inner ring of tentacles around the mouth. Some hydroids may consist of colonies of zooids that serve different purposes, such as defense, reproduction and catching prey. The mesoglea of polyps is usually thin and often soft, but that of medusae is usually thick and springy, so that it returns to its original shape after muscles around the edge have contracted to squeeze water out, enabling medusae to swim by a sort of jet propulsion.

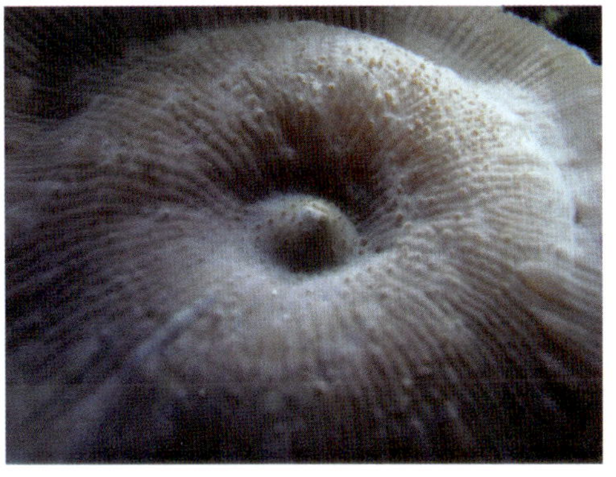

Oral end of actinodiscus polyp, with close-up of the mouth.

Skeletons

In medusae the only supporting structure is the mesoglea. *Hydra* and most sea anemones close their mouths when they are not feeding, and the water in the digestive cavity then acts as a hydrostatic skeleton, rather like a water-filled balloon. Other polyps such as *Tubularia* use columns of water-filled cells for support. Sea pens stiffen the mesoglea with calcium carbonate spicules and tough fibrous proteins, rather like sponges.

In some colonial polyps, a chitinous periderm gives support and some protection to the connecting sections and to the lower parts of individual polyps. Stony corals secrete massive calcium carbonate exoskeletons. A few polyps collect materials such as sand grains and shell fragments, which they attach to their outsides. Some colonial sea anemones stiffen the mesoglea with sediment particles.

Main Cell Layers

Cnidaria are diploblastic animals; in other words, they have two main cell layers, while more complex animals are triploblasts having three main layers. The two main cell layers of cnidarians form epithelia that are mostly one cell thick, and are attached to a fibrous basement membrane, which they secrete. They also secrete the jelly-like mesoglea that separates the layers. The layer that faces outwards, known as the ectoderm ("outside skin"), generally contains the following types of cells:

- Epitheliomuscular cells whose bodies form part of the epithelium but whose bases extend to form muscle fibers in parallel rows. The fibers of the outward-facing cell layer generally run at right angles to the fibers of the inward-facing one. In Anthozoa (anemones, corals, etc.) and Scyphozoa (jellyfish), the mesoglea also contains some muscle cells.

- Cnidocytes, the harpoon-like "nettle cells" that give the phylum Cnidaria its name. These appear between or sometimes on top of the muscle cells.
- Nerve cells. Sensory cells appear between or sometimes on top of the muscle cells, and communicate via synapses (gaps across which chemical signals flow) with motor nerve cells, which lie mostly between the bases of the muscle cells. Some form a simple nerve net.
- Interstitial cells, which are unspecialized and can replace lost or damaged cells by transforming into the appropriate types. These are found between the bases of muscle cells.

In addition to epitheliomuscular, nerve and interstitial cells, the inward-facing gastroderm ("stomach skin") contains gland cells that secrete digestive enzymes. In some species it also contains low concentrations of cnidocytes, which are used to subdue prey that is still struggling.

The mesoglea contains small numbers of amoeba-like cells, and muscle cells in some species. However, the number of middle-layer cells and types are much lower than in sponges.

Polymorphism

Polymorphism refers to the occurrence of structurally and functionally more than two different types of individuals within the same organism. It is a characteristic feature of Cnidarians, particularly the polyp and medusa forms, or of zooids within colonial organisms like those in Hydrozoa. In Hydrozoans, colonial individuals arising from individuals zooids will take on separate tasks. For example, in *Obelia* there are feeding individuals, the gastrozooids; the individuals capable of asexual reproduction only, the gonozooids, blastostyles and free-living or sexually reproducing individuals, the medusae.

Cnidocytes

These "nettle cells" function as harpoons, since their payloads remain connected to the bodies of the cells by threads. Three types of cnidocytes are known:

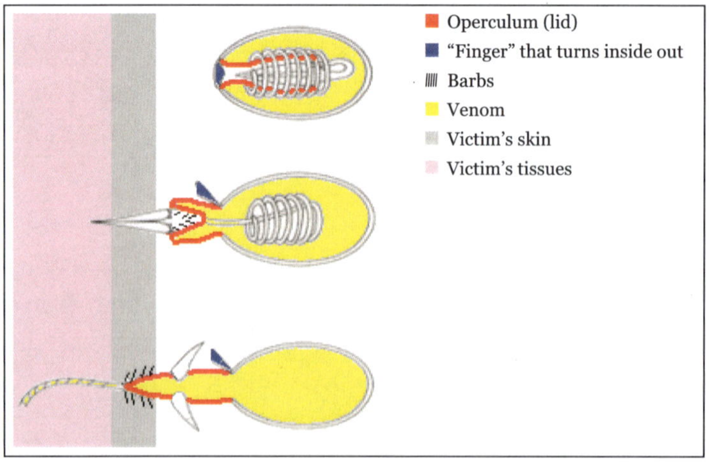

Firing sequence of the cnida in a hydra's nematocyst.

- Nematocysts inject venom into prey, and usually have barbs to keep them embedded in the victims. Most species have nematocysts.

- Spirocysts do not penetrate the victim or inject venom, but entangle it by means of small sticky hairs on the thread.
- Ptychocysts are not used for prey capture — instead the threads of discharged ptychocysts are used for building protective tubes in which their owners live. Ptychocysts are found only in the order Ceriantharia, tube anemones.

The main components of a cnidocyte are:

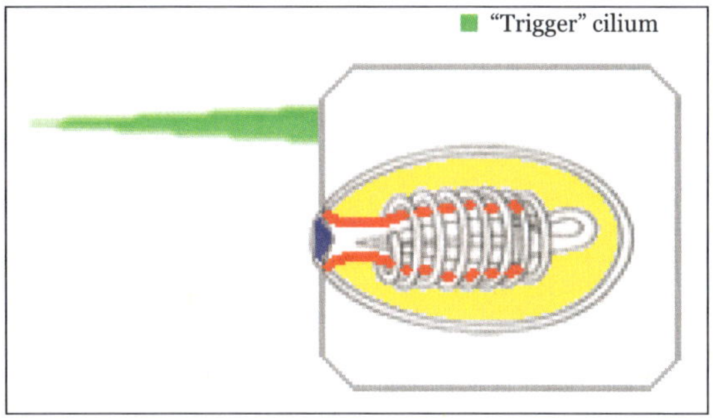

A hydra's nematocyst, before firing.

- A cilium (fine hair) which projects above the surface and acts as a trigger. Spirocysts do not have cilia.
- A tough capsule, the cnida, which houses the thread, its payload and a mixture of chemicals that may include venom or adhesives or both. ("cnida" is derived from the Greek word κνίδη, which means "nettle").
- A tube-like extension of the wall of the cnida that points into the cnida, like the finger of a rubber glove pushed inwards. When a cnidocyte fires, the finger pops out. If the cell is a venomous nematocyte, the "finger"'s tip reveals a set of barbs that anchor it in the prey.
- The thread, which is an extension of the "finger" and coils round it until the cnidocyte fires. The thread is usually hollow and delivers chemicals from the cnida to the target.
- An operculum (lid) over the end of the cnida. The lid may be a single hinged flap or three flaps arranged like slices of pie.
- The cell body, which produces all the other parts.

It is difficult to study the firing mechanisms of cnidocytes as these structures are small but very complex. At least four hypotheses have been proposed:

- Rapid contraction of fibers round the cnida may increase its internal pressure.
- The thread may be like a coiled spring that extends rapidly when released.
- In the case of Chironex (the "sea wasp"), chemical changes in the cnida's contents may cause them to expand rapidly by polymerization.
- Chemical changes in the liquid in the cnida make it a much more concentrated solution,

so that osmotic pressure forces water in very rapidly to dilute it. This mechanism has been observed in nematocysts of the class Hydrozoa, sometimes producing pressures as high as 140 atmospheres, similar to that of scuba air tanks, and fully extending the thread in as little as 2 milliseconds (0.002 second).

Cnidocytes can only fire once, and about 25% of a hydra's nematocysts are lost from its tentacles when capturing a brine shrimp. Used cnidocytes have to be replaced, which takes about 48 hours. To minimise wasteful firing, two types of stimulus are generally required to trigger cnidocytes: nearby sensory cells detect chemicals in the water, and their cilia respond to contact. This combination prevents them from firing at distant or non-living objects. Groups of cnidocytes are usually connected by nerves and, if one fires, the rest of the group requires a weaker minimum stimulus than the cells that fire first.

Locomotion

Medusae swim by a form of jet propulsion: muscles, especially inside the rim of the bell, squeeze water out of the cavity inside the bell, and the springiness of the mesoglea powers the recovery stroke. Since the tissue layers are very thin, they provide too little power to swim against currents and just enough to control movement within currents.

Hydras and some sea anemones can move slowly over rocks and sea or stream beds by various means: creeping like snails, crawling like inchworms, or by somersaulting. A few can swim clumsily by waggling their bases.

Nervous System and Senses

Cnidarians are generally thought to have no brains or even central nervous systems. However, they do have integrative areas of neural tissue that could be considered some form of centralization. Most of their bodies are innervated by decentralized nerve nets that control their swimming musculature and connect with sensory structures, though each clade has slightly different structures. These sensory structures, usually called rhopalia, can generate signals in response to various types of stimuli such as light, pressure, and much more. Medusa usually have several of them around the margin of the bell that work together to control the motor nerve net, that directly innervates the swimming muscles. Most Cnidarians also have a parallel system. In scyphozoans, this takes the form of a diffuse nerve net, which has modulatory effects on the nervous system. As well as forming the "signal cables" between sensory neurons and motoneurons, intermediate neurons in the nerve net can also form ganglia that act as local coordination centers. Communication between nerve cells can occur by chemical synapses or gap junctions in hydrozoans, though gap junctions are not present in all groups. Cnidarians have many of the same neurotransmitters as many animals, including chemicals such as glutamate, GABA, and acetylcholine.

This structure ensures that the musculature is excited rapidly and simultaneously, and can be directly stimulated from any point on the body, and it also is better able to recover after injury.

Medusae and complex swimming colonies such as siphonophores and chondrophores sense tilt and acceleration by means of statocysts, chambers lined with hairs which detect the movements of internal mineral grains called statoliths. If the body tilts in the wrong direction, the animal rights itself by increasing the strength of the swimming movements on the side that is too low. Most species have ocelli ("simple eyes"), which can detect sources of light. However, the agile box jellyfish are unique among Medusae because they possess four kinds of true eyes that have retinas, corneas and lenses. Although the eyes probably do not form images, Cubozoa can clearly distinguish the direction from which light is coming as well as negotiate around solid-colored objects.

Feeding and Excretion

Cnidarians feed in several ways: predation, absorbing dissolved organic chemicals, filtering food particles out of the water, obtaining nutrients from symbiotic algae within their cells, and parasitism. Most obtain the majority of their food from predation but some, including the corals *Hetroxenia* and *Leptogorgia*, depend almost completely on their endosymbionts and on absorbing dissolved nutrients. Cnidaria give their symbiotic algae carbon dioxide, some nutrients, a place in the sun and protection against predators.

Predatory species use their cnidocytes to poison or entangle prey, and those with venomous nematocysts may start digestion by injecting digestive enzymes. The "smell" of fluids from wounded prey makes the tentacles fold inwards and wipe the prey off into the mouth. In medusae the tentacles round the edge of the bell are often short and most of the prey capture is done by "oral arms", which are extensions of the edge of the mouth and are often frilled and sometimes branched to increase their surface area. Medusae often trap prey or suspended food particles by swimming upwards, spreading their tentacles and oral arms and then sinking. In species for which suspended food particles are important, the tentacles and oral arms often have rows of cilia whose beating creates currents that flow towards the mouth, and some produce nets of mucus to trap particles. Their digestion is both intra and extracellular.

Once the food is in the digestive cavity, gland cells in the gastroderm release enzymes that reduce the prey to slurry, usually within a few hours. This circulates through the digestive cavity and, in colonial cnidarians, through the connecting tunnels, so that gastroderm cells can absorb the nutrients. Absorption may take a few hours, and digestion within the cells may take a few days. The circulation of nutrients is driven by water currents produced by cilia in the gastroderm or by muscular movements or both, so that nutrients reach all parts of the digestive cavity. Nutrients reach the outer cell layer by diffusion or, for animals or zooids such as medusae which have thick mesogleas, are transported by mobile cells in the mesoglea.

Indigestible remains of prey are expelled through the mouth. The main waste product of cells' internal processes is ammonia, which is removed by the external and internal water currents.

Respiration

There are no respiratory organs, and both cell layers absorb oxygen from and expel carbon dioxide into the surrounding water. When the water in the digestive cavity becomes stale it must be replaced, and nutrients that have not been absorbed will be expelled with it. Some Anthozoa have ciliated grooves on their tentacles, allowing them to pump water out of and into the digestive cavity without opening the mouth. This improves respiration after feeding and allows these animals, which use the cavity as a hydrostatic skeleton, to control the water pressure in the cavity without expelling undigested food.

Cnidaria that carry photosynthetic symbionts may have the opposite problem, an excess of oxygen, which may prove toxic. The animals produce large quantities of antioxidants to neutralize the excess oxygen.

Regeneration

All cnidarians can regenerate, allowing them to recover from injury and to reproduce asexually. Medusae have limited ability to regenerate, but polyps can do so from small pieces or even collections of separated cells. This enables corals to recover even after apparently being destroyed by predators.

Anatomy of Echinoderm

Echinoderms evolved from animals with bilateral symmetry. Although adult echinoderms possess pentaradial, or five-sided, symmetry, echinoderm larvae are ciliated, free-swimming organisms that organize in bilateral symmetry which makes them look like embryonic chordates. Later, the left side of the body grows at the expense of the right side, which is eventually absorbed. The left side then grows in a pentaradially symmetric fashion, in which the body is arranged in five parts around a central axis. Within the Asterozoa, there can be a few exceptions from the rule. The starfish genus Leptasterias normally have six arms, although five-armed individuals can occur. Also the Brisingida have six armed species. Amongst the brittle stars, six-armed species such as Ophiothela danae, Ophiactis savignyi and Ophionotus hexactis exists, and Ophiacantha vivipara often has more than six.

Echinoderms exhibit secondary radial symmetry in portions of their body at some stage of life. This, however, is an adaptation to their sessile existence. They developed from other members of the Bilateria and exhibit bilateral symmetry in their larval stage. Many crinoids and some seastars exhibit symmetry in multiples of the basic five, with starfish such as *Labidiaster annulatus* known to possess up to fifty arms, and the sea-lily *Comaster schlegelii* having two hundred.

Skin and Skeleton

A brittle star, *Ophionereis reticulata*.

A sea cucumber from Malaysia.

Starfish exhibit a wide range of colours.

Strongylocentrotus purpuratus, a well-armoured sea urchin.

Crinoid on a coral reef.

Echinoderms have a mesodermal skeleton composed of calcareous plates or ossicles. Each one of these, even the articulating spine of a sea urchin, is composed mineralogically of a crystal of calcite. If solid, these would form a heavy skeleton, so they have a sponge-like porous structure known as stereom. Ossicles may be fused together, as in the test of sea urchins, or may articulate with each other as in the arms of sea stars, brittle stars and crinoids. The ossicles may be flat plates or bear external projections in the form of spines, granules or warts and they are supported by a tough epidermis (skin). Skeletal elements are also deployed in some specialized ways, such as the "Aristotle's lantern" mouthparts of sea urchins used for grinding, the supportive stalks of crinoids and the structural "lime ring" of sea cucumbers.

Despite the robustness of the individual skeletal modules complete skeletons of starfish, brittle stars and crinoids are rare in the fossil record. This is because they quickly disarticulate (disconnect from each other) once the encompassing skin rots away, and in the absence of tissue there is nothing to hold the plates together. The modular construction is a result of the growth system employed by echinoderms, which adds new segments at the centre of the radial limbs, pushing the existing plates outwards and lengthening the arms. Sea urchins on the other hand are often well preserved in chalk beds or limestone. During fossilization, the cavities in the stereom are filled in with calcite that is in crystalline continuity with the surrounding material. On fracturing such rock, distinctive cleavage patterns can be seen and sometimes even the intricate internal and external structure of the test.

The epidermis consists of cells responsible for the support and maintenance of the skeleton, as well as pigment cells, mechanoreceptor cells (which detect motion on the animal's surface), and sometimes gland cells which secrete sticky fluids or even toxins. The varied and often vivid colours of echinoderms are produced by the action of skin pigment cells. These are produced by a variable combination of coloured pigments, such as the dark melanin, red carotinoids, and carotene proteins, which can be blue, green, or violet. These may be light-sensitive, and as a result many echinoderms change appearance completely as night falls. The reaction can happen quickly — the sea urchin *Centrostephanus longispinus* changes from jet black to grey-brown in just fifty minutes when exposed to light.

One characteristic of most echinoderms is a special kind of tissue known as "catch connective tissue". This collagenous material can change its mechanical properties in a few seconds or minutes through nervous control rather than by muscular means. This tissue enables a starfish to change from moving flexibly around the seabed to becoming rigid while prying open a bivalve mollusc or preventing itself from being extracted from a crevice. Similarly, sea urchins can lock their normally mobile spines rigidly as a defensive mechanism when attacked.

The Water Vascular System

Echinoderms possess a unique water vascular system. This is a network of fluid-filled canals derived from the coelom (body cavity) that function in gas exchange, feeding, sensory reception and locomotion. This system varies between different classes of echinoderm but typically opens to the exterior through a sieve-like madreporite on the aboral (upper) surface of the animal. The madreporite is linked to a slender duct, the stone canal, which extends to a ring canal that encircles the mouth or oesophagus. From this, radial canals extend along the arms of asteroids and adjoin the test in the ambulacral areas of echinoids. Short lateral canals branch off the radial canals, each one ending in an ampulla. Part of the ampulla can protrude through a pore (or a pair of pores in sea urchins) to the exterior and is known as a podium or tube feet. The water vascular system assists with the distribution of nutrients throughout the animal's body and is most obviously expressed in the tube feet which can be extended or contracted by the redistribution of fluid between the foot and the internal sac.

The organization of the system is somewhat different in ophiuroids where the madreporite may be on the oral surface and the podia lack suckers. In holothuroids, the podia may be reduced or absent and the madreporite opens into the body cavity so that the circulating liquid is coelomic fluid rather than sea water. The arrangements in crinoids is similar to asteroids but the tube feet lack suckers and are used to pass food particles captured by the arms towards the central mouth. In the asteroids, the same wafting motion is employed to move the animal across the ground. Sea urchins use their feet to prevent the larvae of encrusting organisms from settling on their surfaces; potential settlers are moved to the urchin's mouth and eaten. Some burrowing sea stars extend their elongated dorsal tube feet to the surface of the sand or mud above and use them to absorb oxygen from the water column.

Other Organs

Echinoderms possess a simple digestive system which varies according to the animal's diet. Starfish are mostly carnivorous and have a mouth, oesophagus, two-part stomach, intestine and rectum, with the anus located in the centre of the aboral body surface. With a few exceptions, the members of the order Paxillosida do not possess an anus. In many species of starfish, the large cardiac stomach can be everted and digest food outside the body. In other species, whole food items such as molluscs may be ingested. Brittle stars have a blind gut with no intestine or anus. They have varying diets and expel food waste through their mouth. Sea urchins are herbivores and use their specialised mouthparts to graze, tear and chew algae and sometimes other animal or vegetable material. They have an oesophagus, a large stomach and a rectum with the anus at the apex of the test. Sea cucumbers are mostly detritivores, sorting through the sediment with their buccal tentacles which are modified tube feet. Sand and mud accompanies their food through their simple gut which has a long coiled intestine and a capacious cloaca. Crinoids are passive suspension feeders, catching plankton with their outstretched arms. Boluses of mucus-trapped food are passed to the mouth which is linked to the anus by a loop consisting of a short oesophagus and longer intestine.

The coelomic cavities of echinoderms are complex. Aside from the water vascular system, echinoderms have a haemal coelom (or haemal system, the "haemal" being a misnomer), a perivisceral coelom, a gonadal coelom and often also a perihaemal coelom (or perihaemal system). During development, echinoderm coelom is divided in metacoel, mesocoel and protocoel (also called somatocoel, hydrocoel and axocoel, respectively). The water vascular system, haemal system and perihaemal system form the tubular coelomic system. Echinoderms are an exception having both a coelomic circulatory system (i.e., the water vascular system) and a haemal circulatory system (i.e., the haemal and perihaemal systems).

Haemal and perihaemal systems are derived from the coelom and form an open and reduced circulatory system. This usually consists of a central ring and five radial vessels. There is no true heart and the blood often lacks any respiratory pigment. Gaseous exchange occurs via dermal branchiae or papulae in starfish, genital bursae in brittle stars, peristominal gills in sea urchins and cloacal trees in sea cucumbers. Exchange of gases also takes place through the tube feet. Echinoderms lack specialized excretory (waste disposal) organs and so nitrogenous waste, chiefly in the form of ammonia, diffuses out through the respiratory surfaces.

The coelomic fluid contains the coelomocytes, or immune cells. There are several types of immune cells, which vary among classes and species. All classes possess a type of phagocytic amebocyte, which engulf invading particles and infected cells, aggregate or clot, and may be involved in cytotoxicity. These cells are usually larger and granular, and are suggested to be a main line of defense against potential pathogens. Depending on the class, echinoderms may have spherule cells (for cytotoxicity, inflammation, and anti-bacterial activity), vibratile cells (for coelomic fluid movement and clotting), and crystal cells (potential osmoregulatory cells in sea cucumbers),. The coelomocytes also secrete Anti-Microbial Peptides (AMPs) against bacteria, and have a set of lectins and complement proteins as part of an innate immune system that is still being characterized.

Echinoderms have a simple radial nervous system that consists of a modified nerve net consisting of interconnecting neurons with no central brain, although some do possess ganglia. Nerves radiate from central rings around the mouth into each arm or along the body wall; the branches of these nerves coordinate the movements of the organism and the synchronisation of the tube feet. Starfish have sensory cells in the epithelium and have simple eyespots and touch-sensitive tentacle-like tube feet at the tips of their arms. Sea urchins have no particular sense organs but do have statocysts that assist in gravitational orientation, and they have sensory cells in their epidermis, particularly in the tube feet, spines and pedicellariae. Brittle stars, crinoids and sea cucumbers in general do not have sensory organs but some burrowing sea cucumbers of the order Apodida have a single statocyst adjoining each radial nerve and some have an eyespot at the base of each tentacle.

The gonads occupy much of the body cavities of sea urchins and sea cucumbers, while the less voluminous crinoids, brittle stars and starfish have two gonads in each arm. While the ancestral condition is considered to be the possession of one genital aperture, many organisms have multiple gonopores through which eggs or sperm may be released.

Regeneration

Sunflower star regenerating several arms.

Many echinoderms have remarkable powers of regeneration. Many species routinely autotomize and regenerate arms and viscera. Sea cucumbers often discharge parts of their internal organs if they perceive themselves to be threatened. The discharged organs and tissues are regenerated over the course of several months. Sea urchins are constantly replacing spines lost through damage. Sea stars and sea lilies readily lose and regenerate their arms. In most cases, a single severed arm cannot grow into a new starfish in the absence of at least part of the disc. However, in a few species a single arm can survive and develop into a complete individual and in some species, the arms are intentionally detached for the purpose of asexual reproduction. During periods when they have lost their digestive tracts, sea cucumbers live off stored nutrients and absorb dissolved organic matter directly from the water.

The regeneration of lost parts involves both epimorphosis and morphallaxis. In epimorphosis stem cells—either from a reserve pool or those produced by dedifferentiation—form a blastema and generate new tissues. Morphallactic regeneration involves the movement and remodelling of existing tissues to replace lost parts. Direct transdifferentiation of one type of tissue to another during tissue replacement is also observed.

The robust larval growth is responsible for many echinoderms being used as popular model organisms in developmental biology.

Locomotion

Echinoderms primarily use their tube feet to move about but some sea urchins also use their spines. The tube feet typically have a tip shaped like a suction pad in which a vacuum can be created by contraction of muscles. This along with some stickiness provided by the secretion of mucus provides adhesion. Waves of tube feet contractions and relaxations move along the adherent surface and the animal moves slowly along.

Brittle stars are the most agile of the echinoderms, raising their discs and taking strides when moving. The two forward arms grip the substrate with their tube feet, the two side arms "row", the hindermost arm trails and the animal moves in jerks. The arm spines provide traction and when moving among objects, the supple arms can coil around things. A few species creep around on pointed tube feet. Starfish extend their tube feet in the intended direction of travel and grip the substrate by suction, after which the feet are drawn backwards. The movement of multiple tube feet, coordinated in waves, moves the animal forward, but progress is slow. Some burrowing starfish have points rather than suckers on their tube feet and they are able to "glide" across the seabed at a faster rate.

Sea urchins use their tube feet to move around in a similar way to starfish. Some also use their articulated spines to push or lever themselves along or lift their oral surfaces off the substrate. If a sea urchin is overturned, it can extend its tube feet in one ambulacral area far enough to bring them within reach of the substrate and then successively attach feet from the adjoining area until it is righted. Some species bore into rock and they usually do this by grinding away at the surface with their mouthparts.

Sea cucumbers are generally sluggish animals. Many can move on the surface or burrow through sand or mud using peristaltic movements and some have short tube feet on their under surface with which they can creep along in the manner of a starfish. Some species drag themselves along by means of their buccal tentacles while others can expand and contract their body or rhythmically flex it and "swim". Many live in cracks, hollows and burrows and hardly move at all. Some deep water species are pelagic and can float in the water with webbed papillae forming sails or fins.

The majority of crinoids are motile but the sea lilies are sessile and attached to hard substrates by stalks. These stems can bend and the arms can roll and unroll and that is about the limit of the sea lily's movement, although a few species can relocate themselves on the seabed by crawling. The sea feathers are unattached and usually live in crevices, under corals or inside sponges with their arms the only visible part. Some sea feathers emerge at night and perch themselves on nearby eminences to better exploit the food-bearing current. Many species can "walk" across the seabed, raising their body with the help of their arms. Many can also swim with their arms but most are largely sedentary, seldom moving far from their chosen place of concealment.

Feeding

The modes of feeding vary greatly between the different echinoderm taxa. Crinoids and some brittle stars tend to be passive filter-feeders, enmeshing suspended particles from passing water; most sea urchins are grazers, sea cucumbers deposit feeders and the majority of starfish are active hunters.

Crinoids are suspension feeders and spread their arms wide to catch particles floating past. These are caught by the tube feet on the pinnules, moved into the ambulacral grooves, wrapped in mucus and conveyed to the mouth by the cilia lining the grooves. The exact dietary requirements of crinoids have been little researched but in the laboratory they can be fed with diatoms.

Basket stars are suspension feeders, raising their branched arms to collect zooplankton, while brittle stars use several methods of feeding, though usually one predominates. Some are suspension feeders, securing food particles with mucus strands, spines or tube feet on their raised arms. Others are scavengers and feeders on detritus. Others again are voracious carnivores and able to lasso their waterborne prey with a sudden encirclement by their flexible arms. The limbs then bend under the disc to transfer the food to the jaws and mouth.

Many sea urchins feed on algae, often scraping off the thin layer of algae covering the surfaces of rocks with their specialised mouthparts known as Aristotle's lantern. Other species devour smaller organisms, which they may catch with their tube feet. They may also feed on dead fish and other animal matter. Sand dollars may perform suspension feeding and feed on phytoplankton, detritus, algal pieces and the bacterial layer surrounding grains of sand.

Many sea cucumbers are mobile deposit or suspension feeders, using their buccal podia to actively capture food and then stuffing the particles individually into their buccal cavities. Others ingest large quantities of sediment, absorb the organic matter and pass the indigestible mineral particles through their guts. In this way they disturb and process large volumes of substrate, often leaving characteristic ridges of sediment on the seabed. Some sea cucumbers live infaunally in burrows, anterior-end down and anus on the surface, swallowing sediment and passing it through their gut. Other burrowers live anterior-end up and wait for detritus to fall into the entrances of the burrows or rake in debris from the surface nearby with their buccal podia.

Nearly all starfish are detritivores or carnivores, though a few are suspension feeders. Small fish landing on the upper surface may be captured by pedicilaria and dead animal matter may be scavenged but the main prey items are living invertebrates, mostly bivalve molluscs. To feed on one of these, the starfish moves over it, attaches its tube feet and exerts pressure on the valves by arching its back. When a small gap between the valves is formed, the starfish inserts part of its stomach into the prey, excretes digestive enzymes and slowly liquefies the soft body parts. As the adductor muscle of the shellfish relaxes, more stomach is inserted and when digestion is complete, the stomach is returned to its usual position in the starfish with its now liquefied bivalve meal inside it. The same everted stomach process is used by other starfish to feed on sponges, sea anemones, corals, detritus and algal films.

Defense Mechanisms

Despite their low nutrition value and the abundance of indigestible calcite, echinoderms are the prey of many organisms, such as crabs, sharks, sea birds and other echinoderms. Defensive strategies employed include the presence of spines, toxins, which can be inherent or delivered through the tube feet, and the discharge of sticky entangling threads by sea cucumbers. Although most echinoderm spines are blunt, those of the crown-of-thorns starfish are long and sharp and can cause a painful puncture wound as the epithelium covering them contains a toxin. Because of their catch connective tissue, which can change rapidly from a flaccid to a rigid state, echinoderms are very difficult to dislodge from crevices. Certain sea cucumbers have a cluster of cuvierian tubules which can be ejected as long sticky threads from their anus and entangle and permanently disable an attacker. Another defensive strategy sometimes adopted by sea cucumbers is to rupture the body wall and discharge the gut and internal organs. The animal has a great regenerative capacity and will regrow the lost parts later. Starfish and brittle stars may undergo autotomy when attacked, an arm becoming detached which may distract the predator for long enough for the animal to escape. Some starfish species can "swim" away from what may be danger, foregoing the regrowth by not losing limbs. It is not unusual to find starfish with arms of different sizes in various stages of regrowth.

Anatomy of Platyhelminthes

Anatomy

Flatworms are dorso-ventrally flattened. The epidermis is generally ciliated in the turbellarians, while trematodes and cestodes are covered with a cuticle. Beneath the outer covering are two layers of muscle, an outer circular layer, and an inner longitudinal layer; this arrangement permits an undulating form of locomotion that can be observed in the larger turbellarian species. A saclike digestive cavity, with a single opening to the outside that serves as both mouth and anus, is sometimes present; in the simpler forms it is absent or unbranched, but in higher forms it branches to all parts of the body. The major sense organs, when present, are concentrated in the head, or front end. Although a primitive nerve net is present in some of the simpler forms, others have several nerve cords extending from a brain along the length of the body. The latter pattern of organization is retained in the nervous systems of higher invertebrates, specifically annelids and arthropods.

The reproductive system of flatworms is characteristically hermaphroditic (i.e., each individual produces both eggs and sperm), and cross-fertilization between individuals is typical. While trematodes and cestodes shed eggs almost continuously, turbellarians exhibit seasonal reproductive activity and, in addition, display asexual reproduction and the ability to regenerate severed parts of the body.

All except the simplest flatworms have nephridial tubules, called protonephridia, usually distributed throughout the body. Such structures consist of an external opening and a tubule that branches internally, terminating in a number of blind, bulb-shaped structures called flame bulbs, which bear tufts of cilia. They probably function as excretory and osmoregulatory organs.

Physiological Processes of Flatworms

Free-living species of flatworms are predators or scavengers, whereas parasitic forms feed from the tissues of their hosts. Most flatworms have an incomplete digestive system with an opening, the "mouth," that is also used to expel digestive system wastes. Some species also have an anal opening. The gut may be a simple sac or highly branched. Digestion is extracellular, with enzymes secreted into the space by cells lining the tract, and digested materials taken into the same cells by phagocytosis. One group, the cestodes, does not have a digestive system, because their parasitic lifestyle and the environment in which they live (suspended within the digestive cavity of their host) allows them to absorb nutrients directly across their body wall. Flatworms have an excretory system with a network of tubules throughout the body that open to the environment and nearby flame cells, whose cilia beat to direct waste fluids concentrated in the tubules out of the body. The system is responsible for regulation of dissolved salts and excretion of nitrogenous wastes. The nervous system consists of a pair of nerve cords running the length of the body with connections between them and a large ganglion or concentration of nerve cells at the anterior end of the worm; here, there may also be a concentration of photosensory and chemosensory cells.

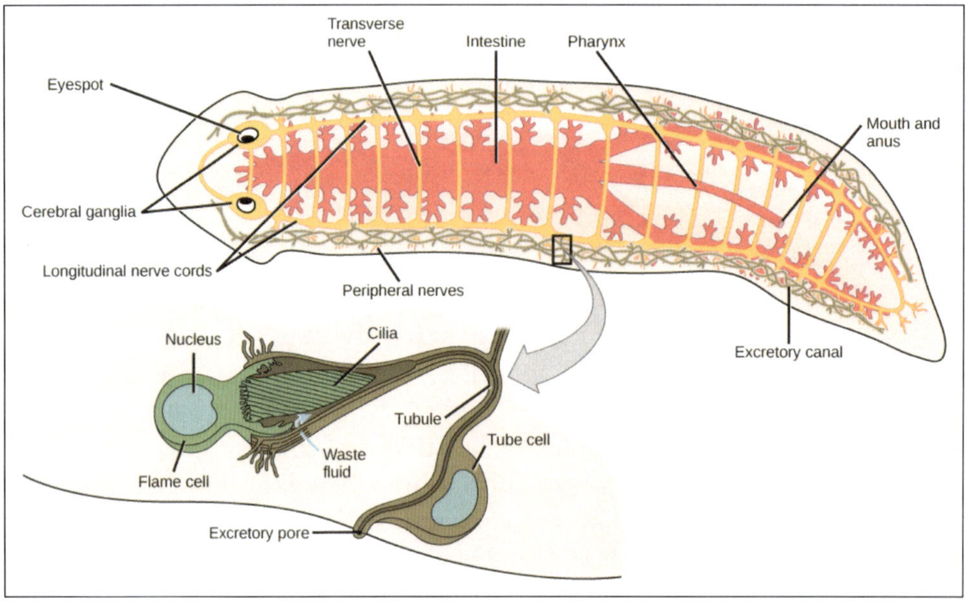

This planarian is a free-living flatworm that has an incomplete digestive system, an excretory system with a network of tubules throughout the body, and a nervous system made up of nerve cords running the length of the body with a concentration of nerves and photosensory and chemosensory cells at the anterior end.

Since there is no circulatory or respiratory system, gas and nutrient exchange is dependent on diffusion and intercellular junctions. This necessarily limits the thickness of the body in these organisms, constraining them to be "flat" worms. Most flatworm species are monoecious (hermaphroditic, possessing both sets of sex organs), and fertilization is typically internal. Asexual reproduction is common in some groups in which an entire organism can be regenerated from just a part of itself.

Anatomy of Nematodes

Nematodes are very small, slender worms: typically about 5 to 100 μm thick, and 0.1 to 2.5 mm long. The smallest nematodes are microscopic, while free-living species can reach as much as 5 cm (2 in), and some parasitic species are larger still, reaching over 1 m (3 ft) in length. The body is often ornamented with ridges, rings, bristles, or other distinctive structures.

The head of a nematode is relatively distinct. Whereas the rest of the body is bilaterally symmetrical, the head is radially symmetrical, with sensory bristles and, in many cases, solid 'head-shields' radiating outwards around the mouth. The mouth has either three or six lips, which often bear a series of teeth on their inner edges. An adhesive 'caudal gland' is often found at the tip of the tail.

The epidermis is either a syncytium or a single layer of cells, and is covered by a thick collagenous cuticle. The cuticle is often of a complex structure and may have two or three distinct layers. Underneath the epidermis lies a layer of longitudinal muscle cells. The relatively rigid cuticle works with the muscles to create a hydroskeleton, as nematodes lack circumferential muscles. Projections run from the inner surface of muscle cells towards the nerve cords; this is a unique arrangement in the animal kingdom, in which nerve cells normally extend fibers into the muscles rather than *vice versa*.

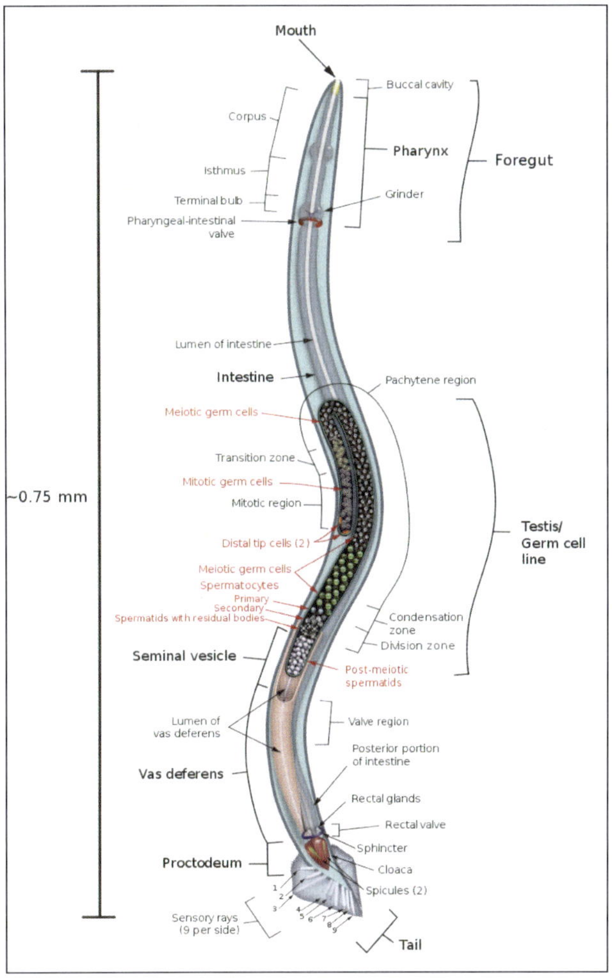

Internal anatomy of a male *C. elegans* nematode.

Digestive System

The oral cavity is lined with cuticle, which is often strengthened with structures, such as ridges, and especially in carnivorous species, which may bear a number of teeth. The mouth often includes a sharp stylet, which the animal can thrust into its prey. In some species, the stylet is hollow and can be used to suck liquids from plants or animals.

The oral cavity opens into a muscular, sucking pharynx, also lined with cuticle. Digestive glands are found in this region of the gut, producing enzymes that start to break down the food. In stylet-bearing species, these may even be injected into the prey.

No stomach is present, with the pharynx connecting directly to a muscleless intestine that forms the main length of the gut. This produces further enzymes, and also absorbs nutrients through its single-cell-thick lining. The last portion of the intestine is lined by cuticle, forming a rectum, which expels waste through the anus just below and in front of the tip of the tail. The movement of food through the digestive system is the result of the body movements of the worm. The intestine has valves or sphincters at either end to help control the movement of food through the body.

Excretory System

Nitrogenous waste is excreted in the form of ammonia through the body wall, and is not associated with any specific organs. However, the structures for excreting salt to maintain osmoregulation are typically more complex.

In many marine nematodes, one or two unicellulars 'renette glands' excrete salt through a pore on the underside of the animal, close to the pharynx. In most other nematodes, these specialized cells have been replaced by an organ consisting of two parallel ducts connected by a single transverse duct. This transverse duct opens into a common canal that runs to the excretory pore.

Nervous System

Four peripheral nerves run the length of the body on the dorsal, ventral, and lateral surfaces. Each nerve lies within a cord of connective tissue lying beneath the cuticle and between the muscle cells. The ventral nerve is the largest, and has a double structure forward of the excretory pore. The dorsal nerve is responsible for motor control, while the lateral nerves are sensory, and the ventral combines both functions.

The nervous system is also the only place in the nematode body that contains cilia, which are all nonmotile and with a sensory function.

At the anterior end of the animal, the nerves branch from a dense, circular nerve (nerve ring) round surrounding the pharynx, and serving as the brain. Smaller nerves run forward from the ring to supply the sensory organs of the head.

The bodies of nematodes are covered in numerous sensory bristles and papillae that together provide a sense of touch. Behind the sensory bristles on the head lie two small pits, or 'amphids'. These are well supplied with nerve cells and are probably chemoreception organs. A few aquatic nematodes possess what appear to be pigmented eye-spots, but whether or not these are actually sensory in nature is unclear.

Anatomy of Annelida

No single feature distinguishes Annelids from other invertebrate phyla, but they have a distinctive combination of features. Their bodies are long, with segments that are divided externally by shallow ring-like constrictions called annuli and internally by septa ("partitions") at the same points, although in some species the septa are incomplete and in a few cases missing. Most of the segments contain the same sets of organs, although sharing a common gut, circulatory system and nervous system makes them inter-dependent. Their bodies are covered by a cuticle (outer covering) that does not contain cells but is secreted by cells in the skin underneath, is made of tough but flexible collagen and does not molt – on the other hand arthropods' cuticles are made of the more rigid α-chitin, and molt until the arthropods reach their full size. Most annelids have closed circulatory systems, where the blood makes its entire circuit via blood vessels.

Summary of distinguishing features						
	Annelida	Recently merged into Annelida		Closely related	Similar-looking phyla	
		Echiura	Sipuncula	Nemertea	Arthropoda	Onychophora
External segmentation	Yes	no	no	Only in a few species	Yes, except in mites	no
Repetition of internal organs	Yes	no	no	Yes	In primitive forms	Yes
Septa between segments	In most species	no	no	No	No	No
Cuticle material	collagen	collagen	collagen	none	α-chitin	α-chitin
Molting	Generally no; but some polychaetes molt their jaws, and leeches molt their skins	no	no	no	Yes	Yes
Body cavity	Coelom; but this is reduced or missing in many leeches and some small polychaetes	two coelomata, main and in proboscis	two coelomata, main and in tentacles	Coelom only in proboscis	Hemocoel	Hemocoel
Circulatory system	Closed in most species	Open outflow, return via branched vein	Open	Closed	Open	Open

Segmentation

Most of an annelid's body consists of segments that are practically identical, having the same sets of internal organs and external chaetae and, in some species, appendages. The frontmost and rearmost sections are not regarded as true segments as they do not contain the standard sets of organs and do not develop in the same way as the true segments. The frontmost section, called the prostomium contains the brain and sense organs, while the rearmost, called the pygidium or periproct contains the anus, generally on the underside. The first section behind the prostomium, called the peristomium, is regarded by some zoologists as not a true segment, but in some polychaetes the peristomium has chetae and appendages like those of other segments.

The segments develop one at a time from a growth zone just ahead of the pygidium, so that an annelid's youngest segment is just in front of the growth zone while the peristomium is the oldest. This pattern is called teloblastic growth. Some groups of annelids, including all leeches, have fixed maximum numbers of segments, while others add segments throughout their lives.

The phylum's name is derived from the Latin word *annelus*, meaning "little ring".

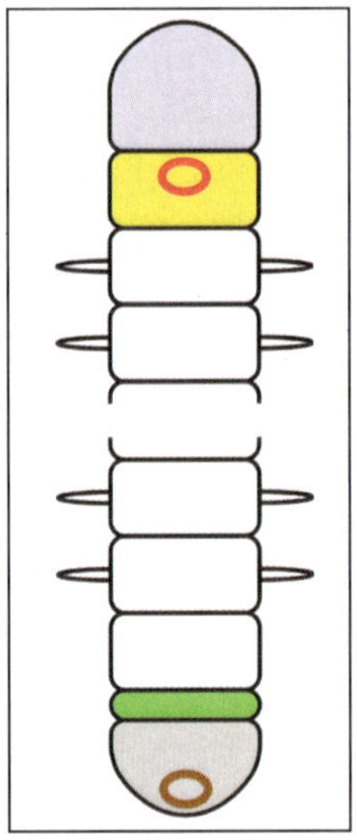

Segments of an annelid.

Body Wall, Chaetae and Parapodia

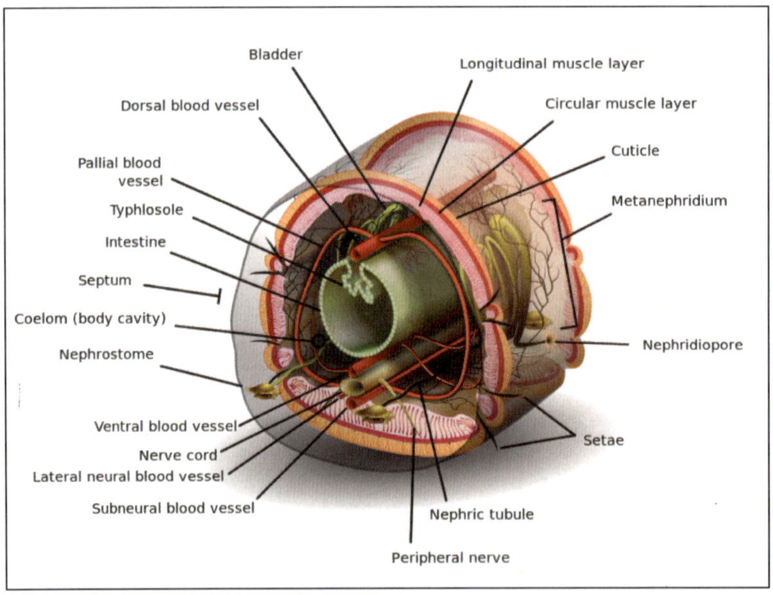

Annelids' cuticles are made of collagen fibers, usually in layers that spiral in alternating directions so that the fibers cross each other. These are secreted by the one-cell deep epidermis (outermost skin layer). A few marine annelids that live in tubes lack cuticles, but their tubes have a similar structure, and mucus-secreting glands in the epidermis protect their skins. Under the epidermis is the dermis, which is made of connective tissue, in other words a combination of cells and non-cellular materials such as collagen. Below this are two layers of muscles, which develop from the lining of the coelom (body cavity): circular muscles make a segment longer and slimmer when they contract, while under them are longitudinal muscles, usually four distinct strips, whose contractions make the segment shorter and fatter. Some annelids also have oblique internal muscles that connect the underside of the body to each side.

The setae ("hairs") of annelids project out from the epidermis to provide traction and other capabilities. The simplest are unjointed and form paired bundles near the top and bottom of each side of each segment. The parapodia ("limbs") of annelids that have them often bear more complex chetae at their tips – for example jointed, comb-like or hooked. Chetae are made of moderately flexible β-chitin and are formed by follicles, each of which has a chetoblast ("hair-forming") cell at the bottom and muscles that can extend or retract the cheta. The chetoblasts produce chetae by forming microvilli, fine hair-like extensions that increase the area available for secreting the cheta. When the cheta is complete, the microvilli withdraw into the chetoblast, leaving parallel tunnels that run almost the full length of the cheta. Hence annelids' chetae are structurally different from the setae ("bristles") of arthropods, which are made of the more rigid α-chitin, have a single internal cavity, and are mounted on flexible joints in shallow pits in the cuticle.

Nearly all polychaetes have parapodia that function as limbs, while other major annelid groups lack them. Parapodia are unjointed paired extensions of the body wall, and their muscles are derived from the circular muscles of the body. They are often supported internally by one or more large, thick chetae. The parapodia of burrowing and tube-dwelling polychaetes are often just ridges whose tips bear hooked chetae. In active crawlers and swimmers the parapodia are often divided into large upper and lower paddles on a very short trunk, and the paddles are generally fringed with chetae and sometimes with cirri (fused bundles of cilia) and gills.

Nervous System and Senses

The brain generally forms a ring round the pharynx (throat), consisting of a pair of ganglia (local control centers) above and in front of the pharynx, linked by nerve cords either side of the pharynx to another pair of ganglia just below and behind it. The brains of polychaetes are generally in the prostomium, while those of clitellates are in the peristomium or sometimes the first segment behind the prostomium. In some very mobile and active polychaetes the brain is enlarged and more complex, with visible hindbrain, midbrain and forebrain sections. The rest of the central nervous system, the ventral nerve cord, is generally "ladder-like", consisting of a pair of nerve cords that run through the bottom part of the body and have in each segment paired ganglia linked by a transverse connection. From each segmental ganglion a branching system of local nerves runs into the body wall and then encircles the body. However, in most polychaetes the two main nerve cords are fused, and in the tube-dwelling genus *Owenia* the single nerve chord has no ganglia and is located in the epidermis.

As in arthropods, each muscle fiber (cell) is controlled by more than one neuron, and the speed and power of the fiber's contractions depends on the combined effects of all its neurons. Vertebrates have a different system, in which one neuron controls a group of muscle fibers. Most annelids' longitudinal nerve trunks include giant axons (the output signal lines of nerve cells). Their large diameter decreases their resistance, which allows them to transmit signals exceptionally fast. This enables these worms to withdraw rapidly from danger by shortening their bodies. Experiments have shown that cutting the giant axons prevents this escape response but does not affect normal movement.

The sensors are primarily single cells that detect light, chemicals, pressure waves and contact, and are present on the head, appendages (if any) and other parts of the body. Nuchal ("on the neck") organs are paired, ciliated structures found only in polychaetes, and are thought to be chemosensors. Some polychaetes also have various combinations of ocelli ("little eyes") that detect the direction from which light is coming and camera eyes or compound eyes that can probably form images. The compound eyes probably evolved independently of arthropods' eyes. Some tube-worms use ocelli widely spread over their bodies to detect the shadows of fish, so that they can quickly withdraw into their tubes. Some burrowing and tube-dwelling polychaetes have statocysts (tilt and balance sensors) that tell them which way is down. A few polychaete genera have on the undersides of their heads palps that are used both in feeding and as "feelers", and some of these also have antennae that are structurally similar but probably are used mainly as "feelers".

Coelom, Locomotion and Circulatory System

Most annelids have a pair of coelomata (body cavities) in each segment, separated from other segments by septa and from each other by vertical mesenteries. Each septum forms a sandwich with connective tissue in the middle and mesothelium (membrane that serves as a lining) from the preceding and following segments on either side. Each mesentery is similar except that the mesothelium is the lining of each of the pair of coelomata, and the blood vessels and, in polychaetes, the main nerve cords are embedded in it. The mesothelium is made of modified epitheliomuscular cells; in other words, their bodies form part of the epithelium but their bases extend to form muscle fibers in the body wall. The mesothelium may also form radial and circular muscles on the septa, and circular muscles around the blood vessels and gut. Parts of the mesothelium, especially on the outside of the gut, may also form chloragogen cells that perform similar functions to the livers of vertebrates: producing and storing glycogen and fat; producing the oxygen-carrier hemoglobin; breaking down proteins; and turning nitrogenous waste products into ammonia and urea to be excreted.

Many annelids move by peristalsis (waves of contraction and expansion that sweep along the body), or flex the body while using parapodia to crawl or swim. In these animals the septa enable the circular and longitudinal muscles to change the shape of individual segments, by making each segment a separate fluid-filled "balloon". However, the septa are often incomplete in annelids that are semi-sessile or that do not move by peristalsis or by movements of parapodia – for example some move by whipping movements of the body, some small marine species move by means of cilia (fine muscle-powered hairs) and some burrowers turn their pharynges (throats) inside out to penetrate the sea-floor and drag themselves into it.

The fluid in the coelomata contains coelomocyte cells that defend the animals against parasites and infections. In some species coelomocytes may also contain a respiratory pigment – red hemoglobin in some species, green chlorocruorin in others (dissolved in the plasma) – and provide oxygen transport within their segments. Respiratory pigment is also dissolved in the blood plasma. Species with well-developed septa generally also have blood vessels running all long their bodies above and below the gut, the upper one carrying blood forwards while the lower one carries it backwards. Networks of capillaries in the body wall and around the gut transfer blood between the main blood vessels and to parts of the segment that need oxygen and nutrients. Both of the major vessels, especially the upper one, can pump blood by contracting. In some annelids the forward end of the upper blood vessel is enlarged with muscles to form a heart, while in the forward ends of many earthworms some of the vessels that connect the upper and lower main vessels function as hearts. Species with poorly developed or no septa generally have no blood vessels and rely on the circulation within the coelom for delivering nutrients and oxygen.

However, leeches and their closest relatives have a body structure that is very uniform within the group but significantly different from that of other annelids, including other members of the Clitellata. In leeches there are no septa, the connective tissue layer of the body wall is so thick that it occupies much of the body, and the two coelomata are widely separated and run the length of the body. They function as the main blood vessels, although they are side-by-side rather than upper and lower. However, they are lined with mesothelium, like the coelomata and unlike the blood vessels of other annelids. Leeches generally use suckers at their front and rear ends to move like inchworms. The anus is on the upper surface of the pygidium.

Respiration

In some annelids, including earthworms, all respiration is via the skin. However, many polychaetes and some clitellates (the group to which earthworms belong) have gills associated with most segments, often as extensions of the parapodia in polychaetes. The gills of tube-dwellers and burrowers usually cluster around whichever end has the stronger water flow.

Feeding and Excretion

Lamellibrachian tube worms have no gut and gain nutrients from chemoautotrophic bacteria living inside them.

Feeding structures in the mouth region vary widely, and have little correlation with the animals' diets. Many polychaetes have a muscular pharynx that can be everted (turned inside out to extend it). In these animals the foremost few segments often lack septa so that, when the muscles in these segments contract, the sharp increase in fluid pressure from all these segments everts the pharynx very quickly. Two families, the Eunicidae and Phyllodocidae, have evolved jaws, which can be used for seizing prey, biting off pieces of vegetation, or grasping dead and decaying matter. On the other hand, some predatory polychaetes have neither jaws nor eversible pharynges. Selective deposit feeders generally live in tubes on the sea-floor and use palps to find food particles in the sediment and then wipe them into their mouths. Filter feeders use "crowns" of palps covered in cilia that wash food particles towards their mouths. Non-selective deposit feeders ingest soil or marine sediments via mouths that are generally unspecialized. Some clitellates have sticky pads in the roofs of their mouths, and some of these can evert the pads to capture prey. Leeches often have an eversible proboscis, or a muscular pharynx with two or three teeth.

The gut is generally an almost straight tube supported by the mesenteries (vertical partitions within segments), and ends with the anus on the underside of the pygidium. However, in members of the tube-dwelling family Siboglinidae the gut is blocked by a swollen lining that houses symbiotic bacteria, which can make up 15% of the worms' total weight. The bacteria convert inorganic matter – such as hydrogen sulfide and carbon dioxide from hydrothermal vents, or methane from seeps – to organic matter that feeds themselves and their hosts, while the worms extend their palps into the gas flows to absorb the gases needed by the bacteria.

Annelids with blood vessels use metanephridia to remove soluble waste products, while those without use protonephridia. Both of these systems use a two-stage filtration process, in which fluid and waste products are first extracted and these are filtered again to re-absorb any re-usable materials while dumping toxic and spent materials as urine. The difference is that protonephridia combine both filtration stages in the same organ, while metanephridia perform only the second filtration and rely on other mechanisms for the first – in annelids special filter cells in the walls of the blood vessels let fluids and other small molecules pass into the coelomic fluid, where it circulates to the metanephridia. In annelids the points at which fluid enters the protonephridia or metanephridia are on the forward side of a septum while the second-stage filter and the nephridiopore (exit opening in the body wall) are in the following segment. As a result, the hindmost segment (before the growth zone and pygidium) has no structure that extracts its wastes, as there is no following segment to filter and discharge them, while the first segment contains an extraction structure that passes wastes to the second, but does not contain the structures that re-filter and discharge urine.

Anatomy of Arthropoda

Segmentation

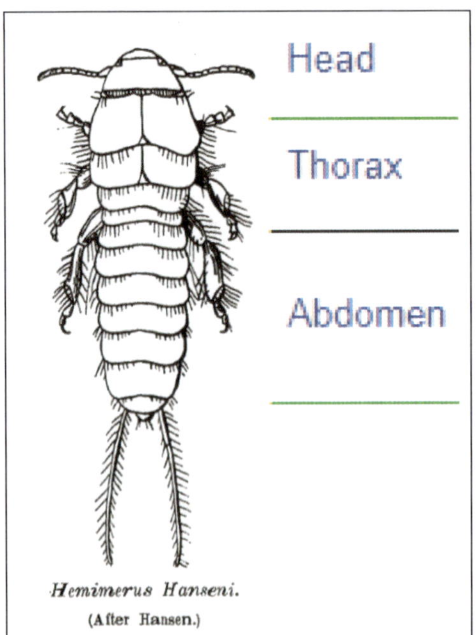

Segments and tagmata of an arthropod.

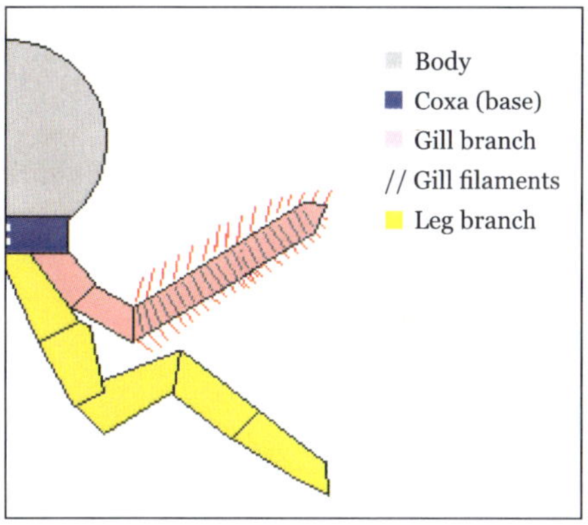

Structure of a biramous appendage.

The embryos of all arthropods are segmented, built from a series of repeated modules. The last common ancestor of living arthropods probably consisted of a series of undifferentiated segments, each with a pair of appendages that functioned as limbs. However, all known living and fossil arthropods have grouped segments into tagmata in which segments and their limbs are specialized in various ways.

The three-part appearance of many insect bodies and the two-part appearance of spiders is a result of this grouping; in fact there are no external signs of segmentation in mites. Arthropods also have two body elements that are not part of this serially repeated pattern of segments, an acron at the front, ahead of the mouth, and a telson at the rear, behind the anus. The eyes are mounted on the acron.

Originally it seems that each appendage-bearing segment had two separate pairs of appendages: an upper and a lower pair. These would later fuse into a single pair of biramous appendages, with the upper branch acting as a gill while the lower branch was used for locomotion. In some segments of all known arthropods the appendages have been modified, for example to form gills, mouth-parts, antennae for collecting information, or claws for grasping; arthropods are "like Swiss Army knives, each equipped with a unique set of specialized tools." In many arthropods, appendages have vanished from some regions of the body; it is particularly common for abdominal appendages to have disappeared or be highly modified.

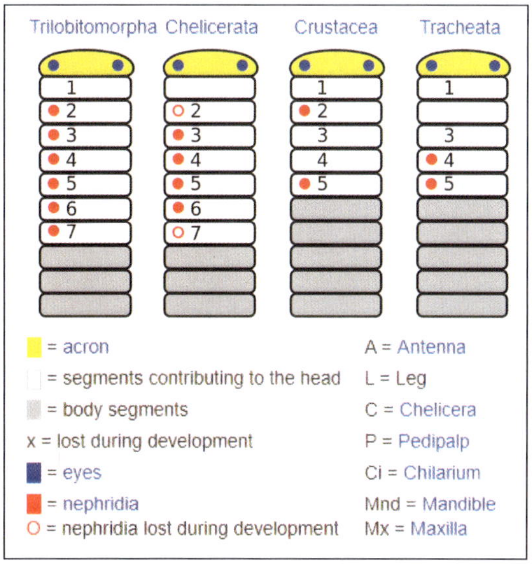

The arthropod head problem.

The most conspicuous specialization of segments is in the head. The four major groups of arthropods – Chelicerata (includes spiders and scorpions), Crustacea (shrimps, lobsters, crabs, etc.), Tracheata (arthropods that breathe via channels into their bodies; includes insects and myriapods), and the extinct trilobites – have heads formed of various combinations of segments, with appendages that are missing or specialized in different ways. In addition, some extinct arthropods, such as *Marrella*, belong to none of these groups, as their heads are formed by their own particular combinations of segments and specialized appendages.

Working out the evolutionary stages by which all these different combinations could have appeared is so difficult that it has long been known as "the arthropod head problem". In 1960, R. E. Snodgrass even hoped it would not be solved, as he found trying to work out solutions to be fun.

Exoskeleton

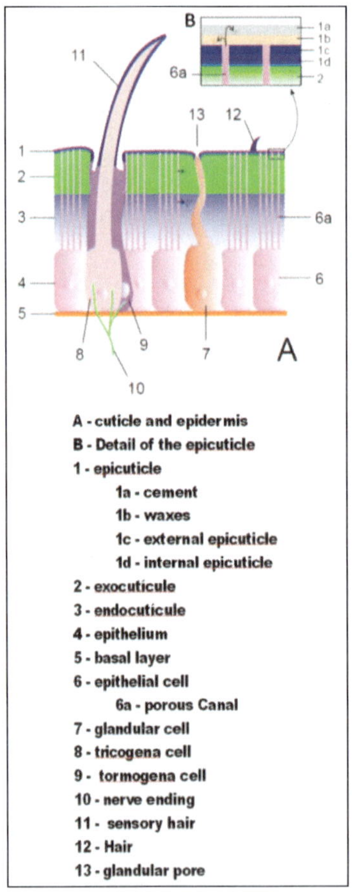

Illustration of an idealized arthropod exoskeleton.

Arthropod exoskeletons are made of cuticle, a non-cellular material secreted by the epidermis. Their cuticles vary in the details of their structure, but generally consist of three main layers: the epicuticle, a thin outer waxy coat that moisture-proofs the other layers and gives them some protection; the exocuticle, which consists of chitin and chemically hardened proteins; and the endocuticle, which consists of chitin and unhardened proteins. The exocuticle and endocuticle together are known as the procuticle. Each body segment and limb section is encased in hardened cuticle. The joints between body segments and between limb sections are covered by flexible cuticle.

The exoskeletons of most aquatic crustaceans are biomineralized with calcium carbonate extracted from the water. Some terrestrial crustaceans have developed means of storing the mineral, since on land they cannot rely on a steady supply of dissolved calcium carbonate. Biomineralization generally affects the exocuticle and the outer part of the endocuticle. Two recent hypotheses about the evolution of biomineralization in arthropods and other groups of animals propose that it provides tougher defensive armor, and that it allows animals to grow larger and stronger by providing more rigid skeletons; and in either case a mineral-organic composite exoskeleton is cheaper to build than an all-organic one of comparable strength.

The cuticle may have setae (bristles) growing from special cells in the epidermis. Setae are as varied in form and function as appendages. For example, they are often used as sensors to detect air or water currents, or contact with objects; aquatic arthropods use feather-like setae to increase the surface area of swimming appendages and to filter food particles out of water; aquatic insects, which are air-breathers, use thick felt-like coats of setae to trap air, extending the time they can spend under water; heavy, rigid setae serve as defensive spines.

Although all arthropods use muscles attached to the inside of the exoskeleton to flex their limbs, some still use hydraulic pressure to extend them, a system inherited from their pre-arthropod ancestors; for example, all spiders extend their legs hydraulically and can generate pressures up to eight times their resting level.

Moulting

Cicada climbing out of its exoskeleton while attached to tree.

The exoskeleton cannot stretch and thus restricts growth. Arthropods therefore replace their exoskeletons by undergoing ecdysis (moulting), or shedding the old exoskeleton after growing a new one that is not yet hardened. Moulting cycles run nearly continuously until an arthropod reaches full size.

In the initial phase of moulting, the animal stops feeding and its epidermis releases moulting fluid, a mixture of enzymes that digests the endocuticle and thus detaches the old cuticle. This phase begins when the epidermis has secreted a new epicuticle to protect it from the enzymes, and the epidermis secretes the new exocuticle while the old cuticle is detaching. When this stage is complete, the animal makes its body swell by taking in a large quantity of water or air, and this makes the old cuticle split along predefined weaknesses where the old exocuticle was thinnest. It commonly takes several minutes for the animal to struggle out of the old cuticle. At this point, the new one is wrinkled and so soft that the animal cannot support itself and finds it very difficult to move, and the new endocuticle has not yet formed. The animal continues to pump itself up to stretch the new cuticle as much as possible, then hardens the new exocuticle and eliminates the excess air or water. By the end of this phase, the new endocuticle has formed. Many arthropods then eat the discarded cuticle to reclaim its materials.

Because arthropods are unprotected and nearly immobilized until the new cuticle has hardened, they are in danger both of being trapped in the old cuticle and of being attacked by predators. Moulting may be responsible for 80 to 90% of all arthropod deaths.

Internal Organs

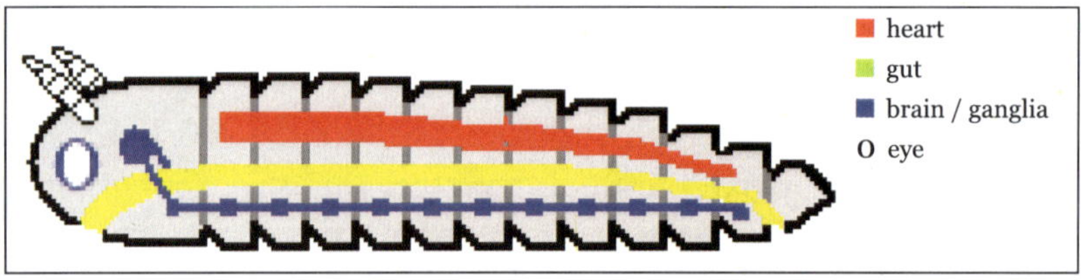

Basic arthropod body structure.

Arthropod bodies are also segmented internally, and the nervous, muscular, circulatory, and excretory systems have repeated components. Arthropods come from a lineage of animals that have a coelom, a membrane-lined cavity between the gut and the body wall that accommodates the internal organs. The strong, segmented limbs of arthropods eliminate the need for one of the coelom's main ancestral functions, as a hydrostatic skeleton, which muscles compress in order to change the animal's shape and thus enable it to move. Hence the coelom of the arthropod is reduced to small areas around the reproductive and excretory systems. Its place is largely taken by a hemocoel, a cavity that runs most of the length of the body and through which blood flows.

Respiration and Circulation

Arthropods have open circulatory systems, although most have a few short, open-ended arteries. In chelicerates and crustaceans, the blood carries oxygen to the tissues, while hexapods use a separate system of tracheae. Many crustaceans, but few chelicerates and tracheates, use respiratory pigments to assist oxygen transport. The most common respiratory pigment in arthropods is copper-based hemocyanin; this is used by many crustaceans and a few centipedes. A few crustaceans and insects use iron-based hemoglobin, the respiratory pigment used by vertebrates. As with other invertebrates, the respiratory pigments of those arthropods that have them are generally dissolved in the blood and rarely enclosed in corpuscles as they are in vertebrates.

The heart is typically a muscular tube that runs just under the back and for most of the length of the hemocoel. It contracts in ripples that run from rear to front, pushing blood forwards. Sections not being squeezed by the heart muscle are expanded either by elastic ligaments or by small muscles, in either case connecting the heart to the body wall. Along the heart run a series of paired ostia, non-return valves that allow blood to enter the heart but prevent it from leaving before it reaches the front.

Arthropods have a wide variety of respiratory systems. Small species often do not have any, since their high ratio of surface area to volume enables simple diffusion through the body surface to supply enough oxygen. Crustacea usually have gills that are modified appendages. Many arachnids have book lungs. Tracheae, systems of branching tunnels that run from the openings in the body walls, deliver oxygen directly to individual cells in many insects, myriapods and arachnids.

Nervous System

Living arthropods have paired main nerve cords running along their bodies below the gut, and in each segment the cords form a pair of ganglia from which sensory and motor nerves run to other parts of the segment. Although the pairs of ganglia in each segment often appear physically fused, they are connected by commissures (relatively large bundles of nerves), which give arthropod nervous systems a characteristic "ladder-like" appearance. The brain is in the head, encircling and mainly *above* the esophagus. It consists of the fused ganglia of the acron and one or two of the foremost segments that form the head – a total of three pairs of ganglia in most arthropods, but only two in chelicerates, which do not have antennae or the ganglion connected to them. The ganglia of other head segments are often close to the brain and function as part of it. In insects these other head ganglia combine into a pair of subesophageal ganglia, under and behind the esophagus. Spiders take this process a step further, as *all* the segmental ganglia are incorporated into the subesophageal ganglia, which occupy most of the space in the cephalothorax (front "super-segment").

Excretory System

There are two different types of arthropod excretory systems. In aquatic arthropods, the end-product of biochemical reactions that metabolise nitrogen is ammonia, which is so toxic that it needs to be diluted as much as possible with water. The ammonia is then eliminated via any permeable membrane, mainly through the gills. All crustaceans use this system, and its high consumption of water may be responsible for the relative lack of success of crustaceans as land animals. Various groups of terrestrial arthropods have independently developed a different system: the end-product of nitrogen metabolism is uric acid, which can be excreted as dry material; the Malpighian tubule system filters the uric acid and other nitrogenous waste out of the blood in the hemocoel, and dumps these materials into the hindgut, from which they are expelled as feces. Most aquatic arthropods and some terrestrial ones also have organs called nephridia ("little kidneys"), which extract other wastes for excretion as urine.

Senses

Optical

The stiff cuticles of arthropods would block out information about the outside world, except that they are penetrated by many sensors or connections from sensors to the nervous system. In fact, arthropods have modified their cuticles into elaborate arrays of sensors. Various touch sensors, mostly setae, respond to different levels of force, from strong contact to very weak air currents. Chemical sensors provide equivalents of taste and smell, often by means of setae. Pressure sensors often take the form of membranes that function as eardrums, but are connected directly to nerves rather than to auditory ossicles. The antennae of most hexapods include sensor packages that monitor humidity, moisture and temperature.

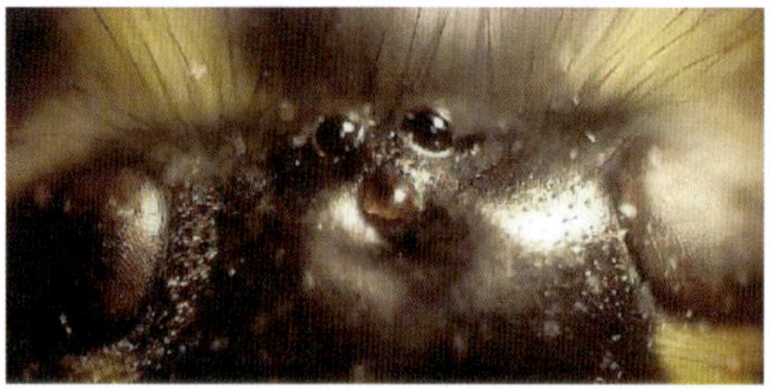

Head of a wasp with three ocelli (center), and compound eyes at the left and right.

Most arthropods have sophisticated visual systems that include one or more usually both of compound eyes and pigment-cup ocelli ("little eyes"). In most cases ocelli are only capable of detecting the direction from which light is coming, using the shadow cast by the walls of the cup. However, the main eyes of spiders are pigment-cup ocelli that are capable of forming images, and those of jumping spiders can rotate to track prey.

Compound eyes consist of fifteen to several thousand independent ommatidia, columns that are usually hexagonal in cross section. Each ommatidium is an independent sensor, with its own light-sensitive cells and often with its own lens and cornea. Compound eyes have a wide field of view, and can detect fast movement and, in some cases, the polarization of light. On the other hand, the relatively large size of ommatidia makes the images rather coarse, and compound eyes are shorter-sighted than those of birds and mammals – although this is not a severe disadvantage, as objects and events within 20 centimetres (7.9 in) are most important to most arthropods. Several arthropods have color vision, and that of some insects has been studied in detail; for example, the ommatidia of bees contain receptors for both green and ultra-violet.

Most arthropods lack balance and acceleration sensors, and rely on their eyes to tell them which way is up. The self-righting behavior of cockroaches is triggered when pressure sensors on the underside of the feet report no pressure. However, many malacostracan crustaceans have statocysts, which provide the same sort of information as the balance and motion sensors of the vertebrate inner ear.

The proprioceptors of arthropods, sensors that report the force exerted by muscles and the degree of bending in the body and joints, are well understood. However, little is known about what other internal sensors arthropods may have.

Anatomy of Mollusca

Muscles and Tissues

The internal molluscan organization is almost entirely soft-bodied. The body cavity is filled with fibrous tissue or fluid-filled spaces (hemocoel), or both. When filled with fluid, the hemocoel expands against the body wall and fibrous tissues, providing a rigid framework and stretching opposing muscles. This same fluid pressure, generated by contraction of other muscles, allows the foot to extend from the shell and penetrate the sediment for burrowing. Conversely, extrusion of the head and foot from the shell in gastropods and cephalopods, shell elevation in gastropods, and the rapid expansion and contraction of the mantle required for jet propulsion in squid and other cephalopods are the result of muscle contractions in the mantle tissue.

The basic sets of muscle systems, fully retained only in solenogasters, include the subintegumental musculature below the mantle; a pair of longitudinal muscle bundles below the mantle margins, which roll the body up and which are almost disintegrated in conchiferans; and the dorsoventral musculature, which is reduced in caudofoveates and shell-less gastropods and which in shelled gastropods forms the columellar muscles that attach the animal to its shell.

The Nervous System and Organs of Sensation

In the nervous system typical of mollusks, a pair of cerebral ganglia (masses of nerve cell bodies) innervate the head, mouth, and associated sense organs. From the dorsal cerebral ganglia, two pairs of longitudinal nerve cords arise: a pair of lateral (pleural) nerve cords, often forming pleural ganglia (which innervate the mantle), and a ventral pair of pedal nerve cords, often forming pedal ganglia (which innervate the foot). In primitive forms both cords are interconnected by lateral branches of nerve fibres. A buccal nerve loop with paired ganglia generally supplies the radular apparatus in the head. Posterior paired visceral ganglia, when present, innervate the viscera. Other mollusks have various grades of ganglia, all of which may be concentrated anteriorly. Because of torsion (that is, a twisting of the body during development), special nerve configurations are found in gastropods; in cephalopods a cartilaginous capsule encloses the concentrated mass of ganglia.

Supplied by the most posterior aspect of the lateral nerve cords, a chemoreceptive sense organ (the osphradium) monitors the water currents entering the mantle cavity. This organ has regressed in scaphopods, some cephalopods, and some gastropods. Pluricellular mantle papillae, which penetrate the cuticle, the valves, and the shell in some conchifers, are differentiated in placophores as photoreceptors. Aside from the well-developed, vertebrate-like eyes of cephalopods, there are photoreceptors on the mantle margins of scallops and related bivalves. Orientation in different gastropods is evidenced by reaction to polarized light, which in part serves for homing. Homing in other gastropods and in the chitons that flee from light appears to be performed by chemoreception along their mucus trails.

The Digestive System

The primitive alimentary tract is straight, and the foregut contains glands and chitinized teeth, called the radula, upon a tough membrane or ribbon underlain by a mass of compact tissue as a support and operated by musculature. In bivalves and some other mollusks the whole radular apparatus is reduced or absent. The radula is used to bite, tear, and scrape various food materials. The different structural aspects of the radula in caudofoveates, solenogasters, and gastropods serve in classification. The differentiation of a more flexible radular structure among the primitive archaeogastropods subsequently enabled successful radiation into diverse habitats.

The midgut in caudofoveates (class Aplacophora) divides into a hindgut and a large ventral sac for enzyme production. In contrast, the midgut in placophores and conchifers is subdivided into a slender esophagus with a pair of glandular pouches, a distinct stomach with a pair of digestive glands, and a slender, often looped intestine. In primitive conchifers the stomach is of the so-called style sac type. The esophagus opens into an anterior elaboration of the stomach into which the enzymes from the style sac, an area separated by ridges, also are released; the tapered end of the stomach leads to the intestine. Cilia that line the style sac churn the stomach contents and form a long food-laden mucous mass called a protostyle, which abuts a chitinous area of epithelium in the stomach. Usually found within the style sac is a rod, called the crystalline style. The protostyle or the crystalline style are fully retained in the bivalves and gastropods that subsist on small microorganisms and detritus. The protostyle or crystalline style may vary in form among the bivalves. Digestion in primitive forms appears to have been both intracellular and extracellular, such as is still the case in solenogasters (class Aplacophora), many bivalves, and most gastropods. In advanced levels either intracellular or extracellular digestion appears to be exclusively elaborated—e.g., advanced crystalline style and intracellular.

The Circulatory System

Mollusks possess an open circulatory system in which body fluid (hemolymph) is transported largely within sinuses devoid of distinct epithelial walls. The posteriodorsal heart enclosed in a pericardium typically consists of a ventricle and two posterior auricles. Hemolymph is drained from ctenidia, gills, or other specialized respiratory epithelia into the respective auricles. The ventricle pumps the hemolymph through a middorsal sinus (in solenogasters and scaphopods) or vessel (aorta) into the body tissues. Hemolymph drains from the tissues into the gills, whence it returns to the auricles.

The respiratory pigment is commonly dissolved in the fluid, either as hemoglobin (as is especially the case in bivalves) or more generally as hemocyanin, which contains copper rather than iron; in more-advanced forms, hemoglobin is bound to blood cells. In chitons and monoplacophorans (but not in the caudofoveates and the solenogasters) the heart is also the site of the purifying ultrafiltration, and the waste products are then discharged into the pericardium and via a pair of pericardial outlets modified to excretory organs (emunctoria, such as false kidneys or nephridia).

The Endocrine System

Hormone production is not well documented in mollusks other than gastropods and cephalopods. Antagonistic neurohormonal control of reproductive activity and metabolic processes is performed in the gastropods through cerebral dorsal bodies and lateral lobes or juxtaposed organs and in the cephalopods through optic glands. In some cephalopods, the hormones also effect death by starvation after the mollusk has deposited its eggs or has mated. Neurosecretions by cells outside the nerve cell bodies (ganglia) have been described in gastropods and cephalopods, the released hormones diffusing through the tissues rather than being concentrated in special organs.

Heart rate in mollusks plays a crucial role in many metabolic processes, including excretion; hormones that affect the heart are released from the wall of veins in cephalopods or, in gastropods, from the subesophageal ganglia, the junction between the auricles and the ventricle. Insulin-like hormones shed from gastropods and bivalves by certain midgut cells control the amount of glucogen (a storage form of sugar) kept as a reserve nutrient.

References

- Phylum-porifera, boundless-biology: courses.lumenlearning.com, Retrieved 28 August, 2019
- Platyhelminthes, plants-and-animals-animals-zoology-invertebrates: encyclopedia.com, Retrieved 05 March, 2019
- Telford, M. J.; Lowe, C. J.; Cameron, C. B.; Ortega-Martinez, O.; Aronowicz, J.; Oliveri, P.; Copley, R. R. (2014). "Phylogenomic analysis of echinoderm class relationships supports Asterozoa". Proceedings of the Royal Society B: Biological Sciences. 281 (1786): 20140479. doi:10.1098/rspb.2014.0479. PMC 4046411. PMID 24850925
- Hodda, M (2011). "Phylum Nematoda Cobb, 1932. In: Zhang, Z.-Q. (Ed.) Animal biodiversity: An outline of higher-level classification and survey of taxonomic richness". Zootaxa. 3148: 63–95. doi:10.11646/zootaxa.3148.1.11
- Rouse, G.W. (2002). "Annelida (Segmented Worms)". Encyclopedia of Life Sciences. John Wiley & Sons, Ltd. doi:10.1038/npg.els.0001599. ISBN 978-0470016176
- Features-of-defense, animal/mollusc- 35781: britannica.com, Retrieved 25 June, 2019
- Collins, A.G. (May 2002). "Phylogeny of Medusozoa and the Evolution of Cnidarian Life Cycles" (PDF). Journal of Evolutionary Biology. 15 (3): 418–432. doi:10.1046/j.1420-9101.2002.00403.x. Archived from the original (PDF) on 2006-09-22. Retrieved 2008-11-2

4

Reproduction in Invertebrates

Reproduction is a biological procedure of producing a new individual from their parents. It is classified into sexual and asexual reproduction. In order to completely understand invertebrates, it is necessary to understand their reproduction processes. The following chapter elucidates the reproduction of invertebrates.

Reproduction in Porifera

Sponges reproduce both asexually and sexually and they also possess the power of regeneration due which it is almost impossible to kill a sponge. Their sexual reproduction is similar to higher animals even though their body organization is primitive type.

Asexual Reproduction

Regeneration

All sponges possess a remarkable ability to regenerate lost parts. A piece cut from the body of a sponge is capable of growing into a complete sponge. If a sponge is cut into small pieces and squeezed through a fine silken mesh to separate cells, the separated amoebocytes will reunite and in a few days will develop canals, flagellated chambers and skeleton and grow up into a new sponge. This power of regeneration helps the sponges to repair the damage caused in the harsh environment.

Budding and Branching

In budding, numerous archaeocytes gather near the surface resulting in a small outgrowth on the pinacoderm. The bud thus formed grows outward to produce a small individual, which either remains attached with the parent individual or gets detached and attached to a nearby rock to grow into an independent colony. Stolon of the sponge grows by branching and secondary branching and many small vertical buds grow out of it.

Fission and Fragmentation

In some sponges multiplication takes place by developing a line of fission and throwing off parts of the body which later can develop into a new sponge. Sponges can break into several pieces along several lines of weakness and breaking into fragments that are capable to tide over unfavourable environmental conditions and grow into complete sponges in the following favourable season.

Reduction Bodies

Many fresh water and marine sponges disintegrate in adverse environmental conditions particularly in winter, leaving small rounded balls called reduction bodies. Each body consists of an internal mass of amoebocytes, covered externally by a pinacoderm and spicules. When favourable conditions return, these reduction bodies grow into complete new sponges.

Gemmules

Fresh water sponges such as Spongilla as well as some marine forms such as Ficulina, Suberites, and Tethya possess the remarkable ability to produce specialised bodies called gemmules, which survive during unfavourable conditions and germinate to produce new sponges. For gemmule formation, archaeocytes laden with food material in the form of glycoprotein or lipoprotein get aggregated into a mass. Amoebocytes surround the central mass of archaeocytes and secrete a thick hard chitinous inner layer and an outer membranous layer over it. Scleroblasts secrete amphidisc spicules between the inner and outer membranes. A fully formed gemmule is a small hard ball having a mass of food laden archaeocytes enclosed in a double layered tough envelope with amphidisc spicules in between. There is a small opening the micropyle through which the cells come out during development in favourable conditions.

In autumn fresh water sponges die and disintegrate, leaving behind a large number of gemmules, which remain viable throughout the winter. Same thing happens during summer when water available is low. In favourable conditions with abundance of water the gemmules begin to hatch and their living contents escape through micropyles and develop into new sponges by collecting themselves together.

Sexual Reproduction

Sexual reproduction involves formation of sperms and ova. The sex cells arise either from archaeocytes or choanocytes. Although most sponges are hermaphrodite but cross-fertilization is the rule because eggs and sperms are produced at different times.

Oocytes are produced inside the body and remain inside mesogloea waiting for fertilization. In sperm formation, archaeocyte and trophocyte cells are involved and in demospongiae choanocytes form sperms and leave the body of sponge through osculum in large numbers.

Sperms from water enter the body of another sponge through canal system and reach the flagellate chambers, where choanocytes trap them. Acting as nurse cells, choanocytes transport the sperm body without tail to the mature ova that wait in the mesogloea. The sperm nucleus then fuses with the nucleus of ovum, ensuring internal fertilization.

Larval Development

Early development takes place within maternal sponge body leading to the formation of a larval stage. Cleavage is holoblastic and radial forming different types of blastula and gastrula.

Development in Syconoid Sponges

In syconoid sponges the larva produced is called stomoblastula, since it has a mouth and feeds on nurse cells within mesogloea and grows for a few days. Stomoblastula after growing changes into amphiblastulaby inverting inside out bringing the flagellated cells on the outer surface so that the larva can swim in water. Amphiblastula leaves the sponge body and swims freely in water feeding on micro-organisms.

Gastrulation is by invagination of micromeres, bringing the flagellated cells again inside the body, lining a cavity which later becomes spongocoel. Cells on the outer surface transform into pinacocytes.

Gastrula swims about and settles on a rock with blastopore against the rock and grows to form olynthusstage that looks like a little sponge. An osculum is formed later.

Development in Asconoid and Leuconoid Sponges

In asconoid and leuconoid sponges, the blastula is called coeloblastula as it does not possess a mouth but has a blastocoel and flagella on the surface of the body. This larva escapes from the sponge body and swims about freely in water.

Gastrulation takes place by delamination of the archaeocytes which are located on one end of the blastocoel. The archaeocytes gradually fill the blastocoel completely and the gastrula becomes solid. This solid gastrula is known as stereogastrula, parenchymula or parenchymella, which swims about for some time and then settles on substratum to form olynthus stage. The inner archaeocytes migrate to form pinacocytes on the surface and the outer flagellated cells migrate towards inside to form choanocytes lining the spongocoel.

Development with Rhagon Larva

In Spongilla, the larva is different from parenchymula and it is called rhagon larva, which has a tent-like body with a broad flat base called hypophare and a conical body called spongophare, with a narrow upper end on which is located the osculum. There are flagellated chambers which open to the outside by ostia and into the spongocoel by apopyles. The sedentary larva grows to become adult.

Reproduction in Cnideria

Sexual

Cnidarian sexual reproduction often involves a complex life cycle with both polyp and medusa stages. For example, in Scyphozoa (jellyfish) and Cubozoa (box jellies) a larva swims until it finds a good site, and then becomes a polyp. This grows normally but then absorbs its tentacles and splits horizontally into a series of disks that become juvenile medusae, a process called strobilation. The juveniles swim off and slowly grow to maturity, while the polyp re-grows and may continue strobilating periodically. The adults have gonads in the gastroderm, and these release ova and sperm into the water in the breeding season.

This phenomenon of succession of differently organized generations (one asexually reproducing, sessile polyp, followed by a free-swimming medusa or a sessile polyp that reproduces sexually) is sometimes called "alternation of asexual and sexual phases" or "metagenesis", but should not be confused with the alternation of generations as found in plants.

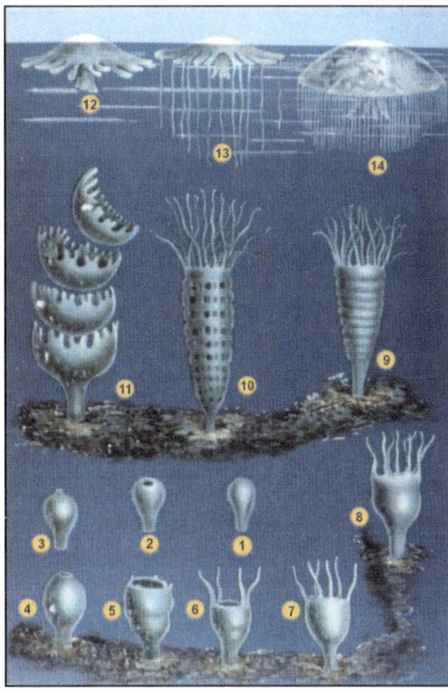

Life cycle of a jellyfish.

Shortened forms of this life cycle are common, for example some oceanic scyphozoans omit the polyp stage completely, and cubozoan polyps produce only one medusa. Hydrozoa have a variety of life cycles. Some have no polyp stages and some (e.g. *hydra*) have no medusae. In some species, the medusae remain attached to the polyp and are responsible for sexual reproduction; in extreme cases these reproductive zooids may not look much like medusae. Meanwhile, life cycle reversal, in which polyps are formed directly from medusae without the involvement of sexual reproduction process, was observed in both Hydrozoa (*Turritopsis dohrnii* and *Laodicea undulata*) and Scyphozoa (*Aurelia* sp.1). Anthozoa have no medusa stage at all and the polyps are responsible for sexual reproduction.

Spawning is generally driven by environmental factors such as changes in the water temperature, and their release is triggered by lighting conditions such as sunrise, sunset or the phase of the moon. Many species of Cnidaria may spawn simultaneously in the same location, so that there are too many ova and sperm for predators to eat more than a tiny percentage – one famous example is the Great Barrier Reef, where at least 110 corals and a few non-cnidarian invertebrates produce enough gametes to turn the water cloudy. These mass spawnings may produce hybrids, some of which can settle and form polyps, but it is not known how long these can survive. In some species the ova release chemicals that attract sperm of the same species.

The fertilized eggs develop into larvae by dividing until there are enough cells to form a hollow sphere (blastula) and then a depression forms at one end (gastrulation) and eventually becomes the digestive cavity. However, in cnidarians the depression forms at the end further from the yolk (at the animal pole), while in bilaterians it forms at the other end (vegetal pole). The larvae, called planulae, swim or crawl by means of cilia. They are cigar-shaped but slightly broader at the "front" end, which is the aboral, vegetal-pole end and eventually attaches to a substrate if the species has a polyp stage.

Anthozoan larvae either have large yolks or are capable of feeding on plankton, and some already have endosymbiotic algae that help to feed them. Since the parents are immobile, these feeding capabilities extend the larvae's range and avoid overcrowding of sites. Scyphozoan and hydrozoan larvae have little yolk and most lack endosymbiotic algae, and therefore have to settle quickly and metamorphose into polyps. Instead, these species rely on their medusae to extend their ranges.

Asexual

All known cnidaria can reproduce asexually by various means, in addition to regenerating after being fragmented. Hydrozoan polyps only bud, while the medusae of some hydrozoans can divide down the middle. Scyphozoan polyps can both bud and split down the middle. In addition to both of these methods, Anthozoa can split horizontally just above the base. Asexual reproduction makes the daughter cnidarian a clone of the adult.

Reproduction in Echinoderm

Sexual Reproduction

Echinoderms become sexually mature after approximately two to three years, depending on the species and the environmental conditions. They are nearly all gonochoric, though a few species are hermaphroditic. The eggs and sperm cells are typically released into open water, where fertilization takes place. The release of sperm and eggs is synchronised in some species, usually with regard to the lunar cycle. In other species, individuals may aggregate during the reproductive season, thereby increasing the likelihood of successful fertilisation. Internal fertilisation has currently been observed in three species of sea star, three brittle stars and a deep water sea cucumber. Even at abyssal depths, where no light penetrates, synchronisation of reproductive activity in echinoderms is surprisingly frequent.

Some echinoderms brood their eggs. This is especially common in cold water species where planktonic larvae might not be able to find sufficient food. These retained eggs are usually few in number and are supplied with large yolks to nourish the developing embryos. In starfish, the female may carry the eggs in special pouches, under her arms, under her arched body or even in her cardiac stomach. Many brittle stars are hermaphrodites. Egg brooding is quite common and usually takes place in special chambers on their oral surfaces, but sometimes the ovary or coelom is used. In these starfish and brittle stars, direct development without passing through a bilateral larval stage usually takes place. A few sea urchins and one species of sand dollar carry their eggs in cavities, or near their anus, holding them in place with their spines. Some sea cucumbers use their buccal tentacles to transfer their eggs to their underside or back where they are retained. In a very small number of species, the eggs are retained in the coelom where they develop viviparously, later emerging through ruptures in the body wall. In some species of crinoid, the embryos develop in special breeding bags, where the eggs are held until sperm released by a male happens to find them.

Asexual Reproduction

One species of seastar, *Ophidiaster granifer*, reproduces asexually by parthenogenesis. In certain other asterozoans, the adults reproduce asexually for a while before they mature after which time they reproduce sexually. In most of these species, asexual reproduction is by transverse fission with the disc splitting in two. Regrowth of both the lost disc area and the missing arms occur so that an individual may have arms of varying lengths. Though in most species at least part of the disc is needed for complete regeneration, in a few species of sea stars, a single severed arm can grow into a complete individual over a period of several months. In at least some of these species, they actively use this as a method of asexual reproduction. A fracture develops on the lower surface of the arm and the arm pulls itself free from the body which holds onto the substrate during the process. During the period of regrowth, they have a few tiny arms and one large arm, thus often being referred to as "comets".

Asexual reproduction by transverse fission has also been observed in adult sea cucumbers. *Holothuria parvula* uses this method frequently, an individual splitting into two a little in front of the midpoint. The two halves each regenerate their missing organs over a period of several months but the missing genital organs are often very slow to develop.

The larvae of some echinoderm species are capable of asexual reproduction. This has long been known to occur among starfish and brittle stars but has been more recently observed in a sea cucumber, a sand dollar and a sea urchin. These species belong to four of the major classes of echinoderms except crinozoans. Asexual reproduction in the planktonic larvae occurs through numerous modes. They may autotomise parts that develop into secondary larvae, grow buds or undergo paratomy. The parts that are autotomised or the buds may develop directly into fully formed larvae or may develop through a gastrula or even a blastula stage. The parts that develop into the new larvae vary from the preoral hood (a mound like structure above the mouth), the side body wall, the postero-lateral arms or their rear ends.

The process of cloning is a cost borne by the larva both in resources as well as in development time. Larvae have been observed to undergo this process when food is plentiful or temperature conditions are optimal. It has also been suggested that cloning may occur to make use of the tissues that are normally lost during metamorphosis. Recent research has shown that the larvae of some sand dollars clone themselves when they detect predators (by sensing dissolved fish mucus). Asexual reproduction produces many smaller larvae that escape better from planktivorous fish.

Reproduction in Platyhelminthes

Both nonparasitic and parasitic flatworms are found in nature. The parasitic worms bear hooks, scolex, suckers and other features for attaching to the host organisms. A classic example of platyhelminthes that most of us are aware of is the tapeworm. It is an intestinal parasite that infests humans and vertebrates in general. Speaking more about phylum platyhelminthes, reproduction takes place by means of asexual and sexual methods. Just like other bodily systems, the platyhelminthes reproductive system is very simple to understand.

Asexual Reproduction

The commonly observed platyhelminthes reproduction is asexual type. Several species of the non-parasitic flatworms undergo transverse fission, in which a single organism splits up into smaller fragments through transverse division. Since flatworms have the ability to regenerate themselves, the lost parts develop and each of the fragment grows into a separate individual. Once, they enter the host organism, they travel to the intestinal tract and start feeding on the partially digested foods. Very soon, they break up into smaller fragments, and each single piece then develops into an individual worm later.

Sexual Reproduction

According to studies conducted on platyhelminthes breeding, it is found that majority of the species are simultaneous hermaphrodites (or monoecious). In other words, both the male and female reproductive organs are present in a single flatworm. This allows a platyhelminth to undergo sexual reproduction on its own. The actual mode of reproduction may differ slightly from one species to another. For all flatworms, fertilization is of internal type, and the male and female gametes fuse inside the body. Some platyhelminthes release cocoons filled with eggs, which later hatch into young worms that are similar to the matured ones.

In other cases of platyhelminthes (e.g., tapeworm), reproduction is preceded by self-fertilization in a single proglottid (segment containing both male and female reproductive structures) or between two proglottids. On the other hand, there are other flatworms that require cross-fertilization for reproduction. Based on the species, fertilized eggs are released from the host organism through feces, or the proglottid filled with eggs is shed as a whole from the host body. The eggs encased in an outer shell remain as they are, until a prospective intermediate host ingests them. Inside the host body, they hatch to larvae and life cycle continues again.

This was an overview concerning 'how do the platyhelminthes undergo reproduction process'. The parasitic flatworms have a specific pattern of reproduction, wherein the sexual stages are completed in the primary (or definitive) host organisms and the asexual stages are present in the intermediate hosts. The rapid multiplication rate of matured worms is a major concern, while dealing with the parasitic forms of flatworms.

Reproduction in Nematodes

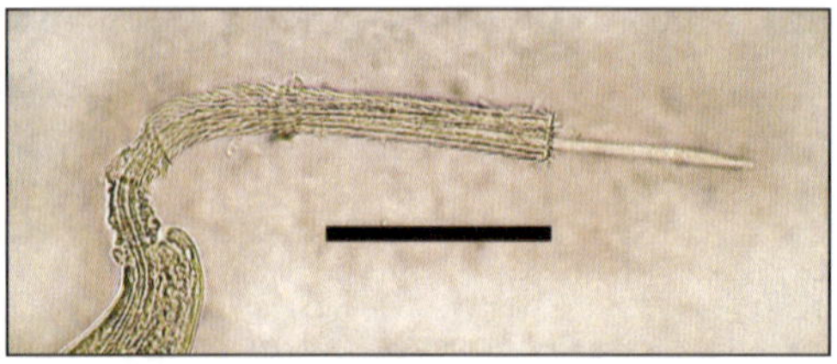

Extremity of a male nematode showing the spicule, used for copulation, bar = 100 μm.

Most nematode species are dioecious, with separate male and female individuals, though some, such as *Caenorhabditis elegans*, are androdioecious, consisting of hermaphrodites and rare males.

Both sexes possess one or two tubular gonads. In males, the sperm are produced at the end of the gonad and migrate along its length as they mature. The testis opens into a relatively wide seminal vesicle and then during intercourse into a glandular and muscular ejaculatory duct associated with the vas deferens and cloaca. In females, the ovaries each open into an oviduct (in hermaphrodites, the eggs enter a spermatheca first) and then a glandular uterus. The uteri both open into a common vulva/vagina, usually located in the middle of the morphologically ventral surface.

Reproduction is usually sexual, though hermaphrodites are capable of self-fertilization. Males are usually smaller than females or hermaphrodites (often much smaller) and often have a characteristically bent or fan-shaped tail. During copulation, one or more chitinized spicules move out of the cloaca and are inserted into the genital pore of the female. Amoeboid sperm crawl along the spicule into the female worm. Nematode sperm is thought to be the only eukaryotic cell without the globular protein G-actin.

Eggs may be embryonated or unembryonated when passed by the female, meaning their fertilized eggs may not yet be developed. A few species are known to be ovoviviparous. The eggs are protected by an outer shell, secreted by the uterus. In free-living roundworms, the eggs hatch into larvae, which appear essentially identical to the adults, except for an underdeveloped reproductive system; in parasitic roundworms, the lifecycle is often much more complicated.

Nematodes as a whole possess a wide range of modes of reproduction. Some nematodes, such as *Heterorhabditis* spp., undergo a process called endotokia matricida: intrauterine birth causing maternal death. Some nematodes are hermaphroditic, and keep their self-fertilized eggs inside the uterus until they hatch. The juvenile nematodes then ingest the parent nematode. This process is significantly promoted in environments with a low food supply.

The nematode model species *C. elegans* and *C. briggsae* exhibit androdioecy, which is very rare among animals. The single genus *Meloidogyne* (root-knot nematodes) exhibits a range of reproductive modes, including sexual reproduction, facultative sexuality (in which most, but not all, generations reproduce asexually), and both meiotic and mitotic parthenogenesis.

The genus *Mesorhabditis* exhibits an unusual form of parthenogenesis, in which sperm-producing males copulate with females, but the sperm do not fuse with the ovum. Contact with the sperm is essential for the ovum to begin dividing, but because no fusion of the cells occurs, the male contributes no genetic material to the offspring, which are essentially clones of the female.

Reproduction in Annelida

Reproduction in annelids is a fairly complex topic. Polychaete and oligochaete species can reproduce both sexually and asexually, while leeches can only reproduce sexually. Asexual reproduction does not involve the formation of gametes (eggs and sperm), and it usually occurs either by budding or fission. During budding, the worm forms a small protrusion, or bud, that slowly develops into a new organism. When the bud has developed to a point where it can function independently of the parent, it breaks off. Fission differs from budding in that the parent simply splits into two pieces, and then each piece reforms the missing components to form a whole organism. Figure below shows a diagram of an annelid in the process of budding.

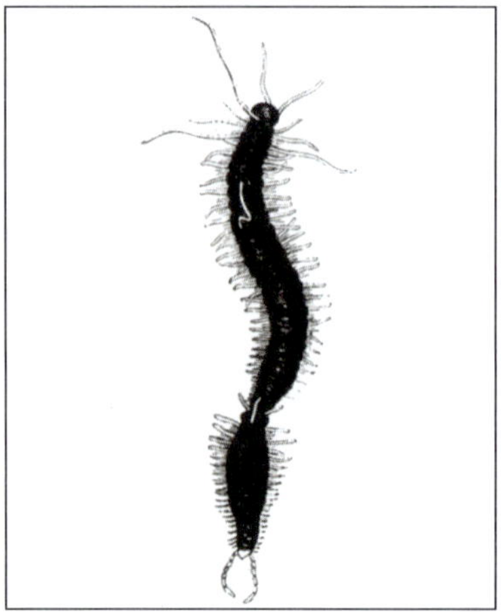

An annelid undergoing asexual reproduction by budding. Notice the new organism forming from the differentiation of segments at the tail end of the animal.

For sexual reproduction involving the formation and fusion of gametes, polychaete speciesnormally have separate sexes, while most oligochaetes are hermaphrodites. In hermaphroditic species, each individual has both male and female sex organs. Sexual reproduction is quite different between polychaetes and oligochaetes.

In order to reproduce sexually, marine polychaetes must undergo a major physical transformation. This transformation varies from species to species, but it generally involves the conversion of several of their posterior segments into specialized reproductive segments containing ovaries or testes that produce gametes. At the same time, organs such as the digestive tract degenerate, and the worm develops enhanced appendages for swimming to the ocean surface. Polychaetes time this reproductive transformation precisely so that many individuals participate at the same time. After they have transformed, they all swim to the surface of the ocean and release their gametes to undergo external fertilization. Following this reproductive burst, the organisms die. The externally fertilized zygotes go on to develop into a specialized larval form called a trochophore larva. This form of reproduction is called epitoky, and there are a number of variations seen in different polychaete species. A polychaete in the epitoky reproductive form is shown in figure below.

A polychaete worm in the reproductive epitoky form. Notice the well-developed appendages for swimming.

Sexual reproduction in oligochaetes and Hirudinea is drastically different from the process of epitoky. Although oligochaetes and Hirudinea are hermaphrodites, they are not able to use their own sperm to self-fertilize their eggs. The hermaphrodites must fertilize each other's eggs through mating. Mating and fertilization are actually separate events in these subclasses. This is possible because the worms have a sac-like organ called a spermatheca that holds onto the sperm after it has been delivered by the mating partner until the worm is ready to use it to fertilize the eggs. Figure below shows two earthworms mating.

Two earthworms mating. In order to mate, the worms line up so that the clitellum of each worm is near the gamete-producing, sexual organs of the other worm. The worms then exchange sperm. The clitella in this picture can be seen as the slightly swollen, pinkish regions on the anterior half of the worm bodies

In order for oligochaetes to fertilize their eggs, they use the clitellum to secrete a cocoon. The clitellum, is a specialized gland-like structure that can be seen from the outside of the animal as a thickened region of segments. The clitellum of each worm can be clearly seen receiving sperm from reproductive segments of its mating partner in figure above.Once the cocoon is ready, they deposit their eggs and sperm, stored from the mating process, into the cocoon where fertilization takes place. In this way, oligochaetes have external fertilization. Leeches carry out a slight variation on this process by using the sperm to fertilize the eggs while they are still within the body. Once they are fertilized, the eggs are then deposited into the cocoon. Therefore, leeches have internal fertilization. Once the fertilized eggs are safely in the cocoon, the worm usually separates from it. The embryos develop and emerge from the cocoon looking like small adults that will continue to grow to full-size. There are no larval stages in the oligochaete and Hirudinea subclasses.

Reproduction in Arthropoda

With few exceptions, the sexes are separate in arthropods; i.e., there are both male and female individuals. The paired sex organs, or gonads, of each sex are connected directly to ducts that open onto the ventral surface of the trunk, the precise location depending upon the arthropod group.

In arthropods, sperm are commonly transferred to the female within sealed packets known as spermatophores. In this method of transfer the sperm are not diluted by the surrounding medium, in the case of aquatic forms, nor do they suffer from rapid desiccation on land. Among some arachnids, such as scorpions, pseudoscorpions, and some mites, the stalked spermatophore is deposited on the ground. Either the female is attracted to the spermatophore chemically or the deposition of the spermatophore occurs during the course of a nuptial dance, and the male afterward maneuvers the female into a position in which she can take up the spermatophore within her genital opening.

Centipedes also utilize spermatophores with an accompanying courtship behavior. Among insects there are some primitive wingless groups, such as collembolans and thysanurans, in which the spermatophore is deposited on the ground, but in most insects the spermatophores are placed directly into the female genital opening by the male during copulation. Many other invertebrates, including several gastropods and chaetognaths, also use spermatophores. Many arthropods transfer free sperm rather than spermatophores. These include many crustaceans, millipedes, some insects (such as dipterans and hemipterans), spiders, and some mites.

Arthropod eggs are usually rich in yolk, but in all groups there are species whose eggs have little yolk. Some specialized methods of reproduction found among certain arthropods include the development of unfertilized eggs (parthenogenesis), the birth of living young (viviparity), and the formation of several embryos from a single fertilized egg (polyembryony).

The eggs of many crustaceans hatch into larvae which have fewer segments than the adult. The earliest larval hatching stage is a minute nauplius larva, which possesses only the first three pairs of appendages. Additional segments and appendages then appear at regular intervals with molting. There are several advantages of larval stages in the development of aquatic animals: Currents disperse the larvae, enabling some to settle in different locations from the parents; because many larvae are capable of feeding, less yolk is required in the egg; and, moreover, planktonic larvae do not compete with benthic adults.

In most chelicerates and insects, almost all of the segments are present at hatching, although in insects the body form may differ from that of the adult. Primitive insects, such as collembolans, have the adult form on hatching. Many insects, such as grasshoppers, crickets, and true bugs, hatch as nymphs, which superficially resemble the adult but lack wings. They gradually acquire these adult features during the nymphal instars. Other insects, such as beetles, butterflies, moths, flies, and wasps, hatch as larvae (grubs, caterpillars, maggots) that differ markedly from the adult. The larvae inhabit different environments and eat different foods than the adults. In these insects a pupal stage with metamorphosis bridges the gap between the larva and the adult form.

Myriapods have the general body form of the adult on hatching though they may lack some of the segments. Most millipedes hatch with only seven trunk segments. Some centipedes hatch with all of the adult trunk segments, but others have fewer than the adult.

The young of most arachnids are similar to the adult. The female scorpion gives birth to her young, which immediately climb onto her back. Female wolf spiders also carry their young, and prior to hatching they carry the white egg case attached to the posterior spinnerets. Unlike other arachnids, mites and ticks hatch as six-legged larvae, which acquire the fourth pair of legs at a later molt.

Reproduction in Mollusca

Mollusks are primarily of separate sexes, and the reproductive organs (gonads) are simple. Reproduction via an unfertilized gamete (parthenogenesis) is also found among gastropods of the subclass Prosobranchia. Most reproduction, however, is by sexual means. Eggs and sperm are released into the water by members of some (primitive) species, and fertilization occurs there.

In prosobranch gastropods, water currents may cause a simple internal fertilization within the mantle cavity, or males may fertilize eggs internally using a muscular penis. Both male and female reproductive organs may be present in one individual (hermaphroditism) in some species, and various groups exhibit different adaptations to this body form. For example, in hermaphroditic bivalves and prosobranch gastropods, male and female gonads are functional at separate times and in rhythmic and consecutive patterns (successive hermaphroditism). Conversely, male and female gonads are functional at the same time (simultaneous hermaphroditism) in solenogasters and many other gastropods.

Fertilization by transfer of capsules containing sperm (spermatophores) typically occurs in cephalopods and some gastropods. In cephalopods, transfer of spermatophores is usually combined with copulation by a modified arm, or hectocotylus. Copulation in solenogasters, often by means of a special genital cone, may be supported by copulatory stylets. Various penis formations, in part with copulatory stylets, or darts, are widely found in gastropods.

Eggs are deposited singly or in groups, generally on some hard surface and often within jelly masses or leathery capsules. Squids of the suborder Oegopsida and some gastropods have eggs that are suspended in the water. Fertilized eggs commonly undergo spiral cleavage, as in annelids and a number of other "protostome" phyla. The eggs of cephalopods, on the other hand, possess a large amount of yolk, which displaces the dividing cells and causes a characteristic type of development.

Many mollusks develop into free-swimming larvae; these larvae are either feeding (planktotrophic) or nonfeeding (lecithotrophic). The larva in primitive bivalves is a pericalymma (test cell) larva in which the embryo is protected below a covering (test) of cells provided with one to four girdles of cilia, at the apex of which is a sensory plate of ciliated cells. After the developing juvenile has grown out apically of the test (which then is lost), the animal settles and develops into an adult. The test in other lecithotrophic larvae is restricted to a preoral girdle of ciliated cells (the prototrochus) and is called the trochophore larva. Trochophores are encountered in the development of many marine annelid species (phylum Annelida). In more advanced mollusks (such as in marine gastropods and bivalves), the trochophore larva develops into a veliger larva. In these generally planktotrophic larvae, the girdle of ciliated cells widens to form a velum that entraps food and also propels the microscopic mollusk through the water. As the larva continues to develop, the shell, mantle cavity, tentacles, and foot appear. After a specific amount of time, which varies according to species and environmental conditions, the larva loses the velum and metamorphoses into an adult. A substantial change in shell morphology usually marks the transition to adult form.

Secondary (newly evolved) larvae have developed among some freshwater bivalves and some cephalopods. Maternal protection of the developing eggs (brood) is not unexceptional behavior in solenogasters, bivalves, and certain gastropod adults. Direct development without a larval stage or the bearing of live young from a yolky egg, or both, are typical in cephalopods and most nonmarine (and many marine) gastropods. Many species go through two breeding seasons per year, whereas in some cephalopod species mating or egg laying appears to be rapidly followed by death effected by hormones.

References

- Reproduction-sponges: iaszoology.com, Retrieved 08 January, 2019
- Platyhelminthes-reproduction: biologywise.com, Retrieved 10 March, 2019
- Annelid-Reproduction-Advanced-BIO-ADV, biology- annelid-reproduction: ck12.org, Retrieved 13 July, 2019
- Evolution-and-paleontology, animal-arthropod: britannica.com, Retrieved 08 February, 2019
- He; et al. (21 December 2015). "Life Cycle Reversal in Aurelia sp.1 (Cnidaria, Scyphozoa)". PLOS ONE. 10 (12): e0145314. Bibcode:2015PLoSO..1045314H. doi:10.1371/journal.pone.0145314. PMC 4687044. PMID 26690755

5

Habitats of Invertebrates

The environment in which organisms live is referred to as habitat. It plays a major role in their life-cycle. Habitat is characterized by physical factors like moisture, soil, temperature, light, etc., and biotic factors like food and predators. This chapter sheds light on the habitat of invertebrates living in pond, marine and terrestrial habitats.

On the basis of Habitat, there are three types of Invertebrates.

Pond Invertebrates

Zooplankton

Zooplankton Are very small to microscopic organisms that live in the water column. Large populations of these creatures can best be seen in clear, green ponds during the summer. Many species of zooplankton migrate daily, spending time in cooler less well lit areas of the pond during the day and rising to the surface at night. Zooplankton often lack pigmentation or other protection from the ultraviolet rays of the sun and would die if continually exposed to strong sunlight. Also, predators are less likely to find zooplankton that remain in darker depths of the pond during the day. In daylight hours most zooplankton will be found in shaded areas of the pond.

A plankton net made to catch these creatures can be purchased from a biological supply house or made from the leg of a pair of pantyhose or women's stockings. Attach one end of the stocking to a wire hoop about 12 inches in diameter. Cut the stocking to about a length of 2 feet and tie a knot in the bottom. To use the net, sweep several times just below the water surface. Evert the stocking over a wide mouth jar and rinse the contents into the jar with a little pond water. Hold the jar of water in the light. Plankton large enough to be visible will be seen as gray to white, pinhead or smaller organisms moving about in the water. They eventually concentrate near the bottom of the container. Zooplankton can be more closely observed by using a magnifying glass or low power microscope. Protozoa and other tiny planktonic organisms require use of a microscope to be seen.

Zooplankton are primary consumers. They feed on phytoplankton (microscopic plants), detritus other zooplankton and bacteria. They are also an important food item for larval fish and other aquatic organisms.

Zooplankton are comprised of a very diverse group of animals including protozoa, rotifers, nematodes, crustaceans and other arthropods. Most of the zooplankton visible to the naked eye or with a small magnifying glass are crustaceans including ostracods, copepods, cladocerans and amphipods.

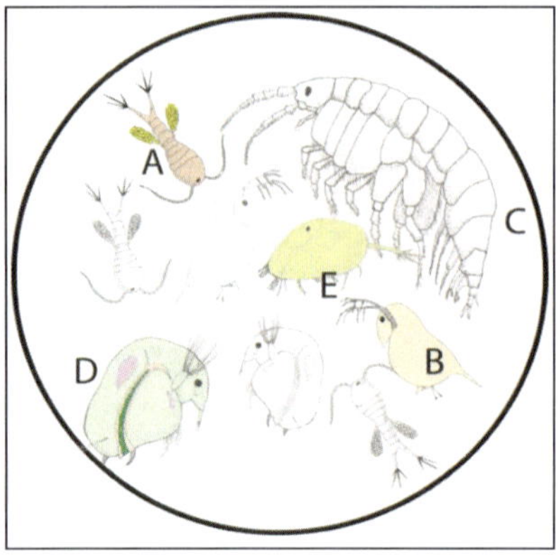

Representative groups of zooplankton. A - copepods, B and D cladocerans, C - amphipod, E - ostrocods.

Fresh-water Sponges

There are about 30 species of fresh-water sponges in the U.S. Usually no more than 2-3 species will be found in the same location. The most common species is Spongilla lacustris. The sponges are highly variable in color, often brown, gray, yellow or greenish when in sunlight. The green color is due to ingested algae that continue to live for awhile and may even multiply before being digested in the body of the sponge.

(a) Spongilla lacustris - Fresh-water sponge on a branch (b) Enlargement showing spicules, oscula and gemmules.

Sponges vary in size from a few cm^2 to over $40m^2$, depending on age and environmental conditions. Living tissue of the sponge is only about 1-4 mm thick but it can look much thicker due to successive layers of new growth accumulating over dead tissue in the same location over a period of several years.

Growth forms may be mat-like or the sponge may have branches or lobes depending on substrate and environmental conditions. Sponges can be found attached to any firm substrate, most commonly, branches, stones, logs and other aquatic debris.

Sponges are not often found in polluted or silt laden waters, or on muddy pond bottoms. Intolerance to water pollution has brought about the extinction of several species.

On close examination with a magnifying glass or low power microscope, the sponge can be seen to have two general sizes of pores on the body surface. Small pores are called ostia and the large pores are called oscula. Flagellated cells in the interior of the sponge create a current that brings water into the sponge through the ostia. The water enters incurrent canals that go to spherical chambers lined with flagellated cells. Food is collected and absorbed by cells lining the chamber called amoebacytes. Water leaves the chamber through the excurrent canal located opposite the incurrent canal. The excurrent canal leads to an expanded chamber that terminates in the osculum, the opening through which water exits the sponge.

The sponge maintains its form by means of silicate structures called spicules. Spicules are often rod shaped and pointed at both ends, however other shapes may also be found in the organism. Spicules are found bundled into columns and also randomly distributed throughout the sponge body.

Sponges feed on microscopic detritus, algae, bacteria and tiny protozoans that enter the sponge body with the water current.

Sponges are preyed upon by crayfish, particularly of the species Orconectes, Spongilla fly larvae and other aquatic insect larvae.

Reproductive structures called gemmules form in the body of the sponge, most often in the autumn but they can form anytime during the year. Gemmules are round, flattened disks, about the size of a pin head, containing the necessary materials to regenerate a new sponge. In cold weather the sponge begins to disintegrate, leaving the gemmules to fall to the bottom or remain in place. In spring the gemmules open through a small pore and the sponge material flows out, covering the gemmule and forming a new sponge.

Sexual reproduction also occurs through the warm months. Individuals may be of both sexes or a single sex. Sperm and eggs are formed in the sponge body and released through the osculum into the environment where they enter other sponges through the ostia. Fertilized eggs develop and migrate to the oscula where they are released and form new sponges.

Hydra

Hydra are interesting and often unnoticed aquatic organisms that can be found in most ponds. When viewed with the unaided eye they appear as white, brown or light green colored threads about ½ inch or less in length. Most hydra are less than a 1/4 inch long when fully extended. When viewed with a magnifying glass or microscope, the tentacles and sometimes, reproductive buds become visible.

Hydra have 4-12 tentacles that are used to capture prey. Zooplankton or microscopic aquatic animals are the main food source of the hydra although they can capture insect larvae, larval fish and tadpoles.

The hydra looks like a flexible hollow tube with tentacles attached at one end. The mouth is located in the center of the ring of tentacles.

The tentacles are covered with wart-like bumps that contain nematocysts or stinging cells. When prey come into contact with the tentacles the nematocysts resembling long threads are released. These threads are sometimes barbed and can pierce the prey.

Chemicals contained in the nematocysts paralyze the prey which is then guided to the mouth by the tentacles. Food is digested in the gastrovascular cavity. Indigestible food is ejected through the mouth.

Tentacles are also used for locomotion. Hydra can move by tumbling from its up right position onto its tentacles and back to the upright position coming to rest on to its basal disc which usually anchors the organism.

When the hydra is disturbed or inactive the tentacles and body shorten. Tentacles and body extend during feeding periods or when the hydra is moving.

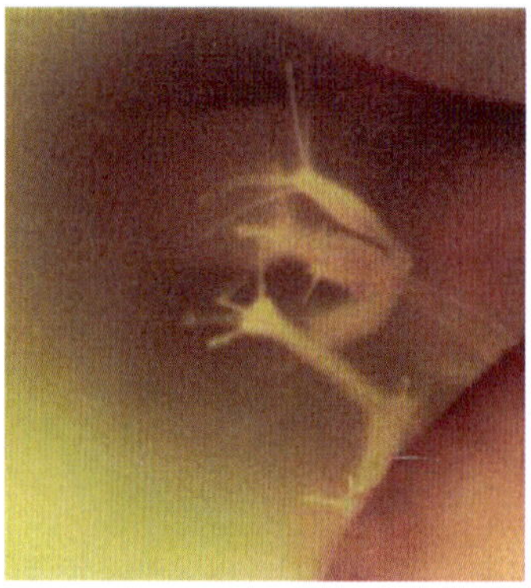

A group of Hydra.

Hydra attached to snail shell.

Collecting Hydra

Hydra can be found attached to the underside of the floating leaves of aquatic plants such as pond weed, water lilies and duckweed. Hydra prefer cool water in shaded locations. They may appear as small jelly-like bubbles or threads on the undersides of leaves to the unaided eye. Places some leaves in a clear glass jar and let it set for a few hours in a cool, shaded location. Observe the sides of the jar with a magnifying glass after the water has settled. Hydra may be found attached to the jar.

If a dissecting microscope or other low power microscope is available, use it to observe the hydra. Place a leaf and some pond water in a petri dish. Use a low power lens to locate the hydra. They may be found attached to the leaf or bottom of the dish. Larger numbers of hydra can be collected by placing about ½ gallon of aquatic vegetation in a 10 gallon aquarium. Fill the aquarium with pond water from the collection site. It will contain sufficient protozoa to feed the hydra for a few days. Aerate the aquarium gently but do not use filtration. Many of the hydra will collect on the sides and floor of the aquarium, others will be seen attached to vegetation in the tank.

Fresh-water Jellyfish

Craspedacusta sowerbyi, the fresh-water jellyfish is related to the hydra. It was not realized that it's hydroid form and medusa form, what is commonly thought of as a jellyfish , was the same animal until 1924.

The hydroid form usually consists of 2-4 individual but connected animals although some hydroid colonies may be comprised of up to 10 hydroids. An individual can break off from the colony and move slowly about the substrate under its own power. A single individual is usually less than 2 mm long. It has no tentacles, however it does have nematocysts or stinging cells around the mouth. Nematocysts sting and immobilize prey. The body of hydroids secretes a sticky mucus. Substrate debris clings to the mucus coat and provides a camouflage covering over the creature. The hydroid feeds on zooplankton and other small organisms living on the pond bottom.

Craspedacusta sowerbyi -fresh-water jellyfish.

Like the hydra, the hydroid form of the fresh-water jellyfish produces buds that break off to form other hydroid forms. At occasional, unpredictable intervals, most often from July through October, a bud producing a medusa will form and break free. Medusa often appear in the water in large numbers and are quite noticeable when they occur.

The young medusa forms are about 0.4 mm in diameter and have 8 tentacles. They grow and mature rapidly, reaching a size up to 22 mm or about the size of a penny. The medusa can swim in all directions using its bell for propulsion.

A mature medusa may have 50-500 tentacles of various lengths. Shorter tentacles are used in feeding. It is believed that long tentacles help stabilize the organism in the water.

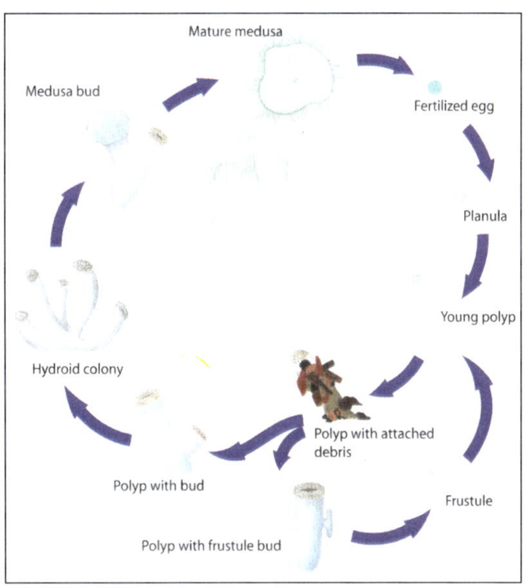

Nematocysts are found on the tentacles, around the mouth and around the edge of the bell. The freshwater jelly fish feeds on zooplankton from about 0.2 - 2 mm in size. It may consume some larval fish but is not considered an important predator. Crayfish are known to eat medusae that are found resting on the pond substrate.

The medusa reaches sexual maturity quickly. Usually, only males or females are observed in a single population. Sperm or eggs are released and fertilization occurs in the water. The fertilized egg develops into a hydroid colony about 2-8 mm in size on the pond bottom.

Fresh-water jelly fish are found in ponds, small lakes and sluggish streams throughout the United States except for the northern New England area.

Flat Worms

Flatworms are a group of organisms that include parasitic tapeworms, flukes and free swimming, non-parasitic turbellarians. Turbellarians may be familiar from their use in biology classes and are often known by a species name, Planaria. Many species exist., most of them less than 4 mm in length. Ones most likely to be encountered in the pond are much larger, 5 - 30 mm in length. They may be one of several species,One of the more common being Dugesia sp.

These ribbon-like worms are colored brown, black, dark yellow, olive green or gray and may be mottled, striped or spotted. The head has two eye-spots that are black and somewhat bean shaped. Ear-like flaps at the sides of the head are used to detect water currents.

The Mouth is found near the center of the body on the ventral side. A feeding tube, the pharynx, can be extended through the mouth and is used to suck up food items.

Turbellarians feed on soft tissues of small invertebrates and dead animal matter. Flatworms are not a significant food item of other organisms in the pond.

Cilia, or microscopic hairs cover the ventral surface of the worm and help it move along the substrate. In some species cilia cover the entire surface of the worm. Flatworms move by secreting a film of mucus onto the substrate and use their cilia to help glide over this substance. Most flatworms cannot freely swim in the water.

Dugesia sp. - flatworms on the stem of a southern niad, an aquatic plant.

Flatworms generally are to be found in shallow water away from direct sunlight. Look for them on the underneath side of leaves, stones or other debris. Flatworms inhabit streams, ponds, springs, marshes and most other bodies of water. They obtain oxygen and release carbon dioxide through their epidermis. Most species require oxygen levels in the water of at least 70% saturation.

Flatworms reproduce both sexually and asexually. Asexual reproduction is by means of fission. The worm's body pinches near the center, gradually splitting into two separate halves that become separate individual worms. Experiments have shown that the worms also can be split longitudinally or cut in half and the separated sections will form new worms. Species that can reproduce by fission have been kept alive up to 11 years in the laboratory.

Sexual reproduction usually requires an exchange of sperm from one worm to another. Both sexes are present in each individual. Up to 20 fertilized eggs are laid in a cocoon that is about 2-4 mm in diameter and brown to black in color. In some species the cocoon is attached to the substrate by a stalk. Young worms are 1-3 mm in length and develop without undergoing any larval stages.

Eggs may over-winter in the cocoon or hatch in about 2 weeks depending on environmental conditions. Flat worms live for a few weeks to months, rarely more than 1 year.

Flat worms can be collected easily. Place a piece of raw meat in a jar. Punch several small holes about 1/8 inch in diameter in the lid of the jar. Cover the jar with duct tape or use a dark colored glass. Place it in a shaded area of the pond in water 6-18 inches deep. Worms may find the bait within an hour but more will be captured if the jar is left over night. The flat worms can be washed off the meat into a dish for closer examination.

Bryozoa

Bryozoa are colonial animals. These groups of organisms can be found from late spring to early fall. They attach themselves to submerged objects such as tree limbs piers, rocks or submerged vegetation.

The more common species form gray, gelatinous, globular shaped colonies that are often round or take on somewhat the shape of the object to which they are attached. Colonies commonly attain the size of a basketball but may be as much as several feet in diameter. The living part of the colony called the zooids are located near the surface of the colonial structure. Tentacled and cilliated filter feeding structures called lophophores extend through an opening in the surface of the colony. The "U" shaped Lophophores retract into a tentacular sheath when not in use. The inside and base of the colony consists of a mass of gelatinous material. Each zooid shares coelomic (body cavity) fluid with other zooids through pores or incomplete cell membranes. Colonies grow best in low light or dark conditions, although enough single cell algae and organic detritus must be present to sustain the organisms.

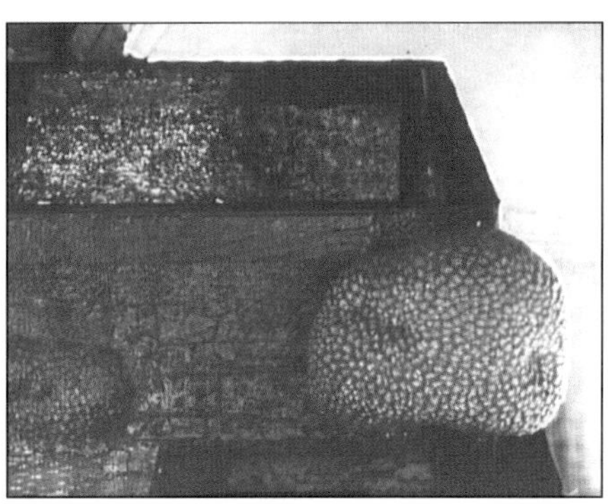

Bryozoa on a fish cage.

Statoblasts, the asexual reproductive phase of freshwater bryozoans, can be found floating in ponds and lakes during summer and fall. Several hundred may be found drifting against rocks or the shoreline. Some drifts of statoblasts can number in the hundreds of thousands. These round gelatinous balls are often puzzling enigmas to those who find them.

Leeches

Leeches are very closely related to the common earth worm. Leeches are usually .5-2 in. long, however, the longest can be up to 18 inches. Leech species are either ectoparasitic and feed on blood of their hosts or are predaceous and feed on small invertebrates such as insect larvae, flatworms, snails and nematodes. Leeches are found in freshwater in temperate regions of the world but can also be found in tropical terrestrial environments that are moist and there also are marine species.

Predaceous species usually swallow prey whole and feed relatively often. The blood sucking leeches must feed only occasionally, 6-9 months between feedings is common. The blood sucking leeches are equipped with 3 semicircular, toothed, cutting plates located in the mouth which is found inside the anterior sucker. The teeth form a triangle inside the mouth.

Glands in the pharynx of the leech excrete an anesthetic that numbs the bitten area. People that have been bitten by leeches report being unaware of the leeches presence on their bodies until they were seen. Glands in the pharynx also secrete an anticoagulant called hirudin which prevents blood from clotting. The hirudin causes leech wounds to bleed profusely until the anticoagulant is washed out of the wound. Leech bites can continue to bleed up to 6 hours after the animal has been removed.

Bloodsucking leeches can take in a large quantity of blood. Many species can consume 6-10 times their own body weight at a single feeding which lasts about 45 minutes. Blood is digested in the leech very slowly , up to 9 months are needed after ingestion to completely digest the meal. Unlike most organisms that rely on secreted acids and enzymes to digest food, many leeches have specific, symbiotic bacteria in the digestive tract that excrete enzymes that digest the blood. These bacteria prevent other bacteria from growing in the gut and help preserve the blood from decomposition during the long digestive period.

Leech On Small pebble.

Leeches in Medicine

Leeches have a long history of use in medicine stretching back at least to the Greek, Nicander of Colophon (c 130 BC). Bleeding patients with leeches was once thought to remove disease and excessive "humors" from the blood. It is estimated that about 100,000 leeches per year are currently used in many hospitals for research and for certain medical procedures. The anticoagulant and vasodilators in the leech saliva can help maintain circulation in small capillaries during reattachment surgery and tissue grafts; opening clogged capillaries and veins. This effect can last for up to 48 hours after the leech detaches.

Finding and Observing Leeches

Most freshwater leeches prefer shallow water bordering ponds and lakes or sluggish streams. Many species are nocturnal. They feed at night and often hide during the day. The best places to search for leeches are under rocks and logs in shallow water during spring and summer. Some species of leeches prefer to feed mostly on one species or class of organism such as frogs, turtles or fish. Other leeches are opportunistic and will feed on whatever is available.

Examine captured vertebrate specimens and look for leeches on the soft tissues.

Some species of leeches can be captured by tying a piece of liver to a string and tossing it into 1-2 feet of water near the shore of a pond. This is best done in the evening or at night. Check the bait periodically to remove crawfish or turtles. Examine the liver for leeches about every 15 minutes because leeches will drop off the liver when they have finished their meal.

Crayfish

Crayfish, also known as crawfish, crawdads or mud bugs, are easy to catch in many ponds. Many people have become familiar with these crustaceans because of their availability in numerous seafood and Cajun style restaurants. The tail is usually the portion of the crawfish that is eaten, although many people "suck the heads" or eat the internal organs as well. Large "pinchers" also contain a portion of meat. Crayfish taste much like lobster or shrimp.

The two most common pond crayfish are the northern crayfish, Orconectes virilis and the red swamp crayfish, Procambarus clarkii.

Coloration is variable among and within species and may be red, brown, blue, green, orange or mottled in appearance.

Crawfish pond.

The body consists of a fused head and thorax called the cephalothorax. The hard outer covering is the carapace. Feathery gills are located under the carapace just above the first joints of the legs. Internal organs are located in the body cavity within the carapace.

Procambarus clarkii - Red swamp crayfish.

Appendages of the crayfish are specialized for different tasks. The first sets of appendages are the two sets of antennae located at the anterior end of the head. Antennae are use for finding food and sensing the environment.

The next set of appendages comprise the mouthparts used for mincing and tearing small bits of food. The claws or chela are the first legs. Chelae are used in defensive and offensive behaviors as well as for grasping and crushing food. The second and third sets of legs have small pinchers used in handling and mincing food items. These legs are also used in walking. The 4th and 5th set of legs are used primarily for walking.

Appendages of the abdomen called pleopods are used for egg attachment sites in the female. The first two pair of male pleopods are modified for sperm transfer.

The terminal appendages are the uropods and telson. Uropods are located on either side of the telson and this fan - shaped appendage is used in locomotion.

Crayfish are most active from dusk to dawn. Many species do not venture from their burrows until the sunlight is reduced from cloudiness or night. These creatures are very mobile. They walk about on the pond bottom or through vegetation. Crayfish are usually found in water less than 1 m deep. If disturbed, crayfish can quickly flex their abdomen and with the aid of the tail dart away backwards.

Crayfish mainly eat dead vegetation and the plants and animals that inhabit this detrital material. They also eat fresh aquatic vegetation, aquatic insects, small fish, snails and commercial fish rations.

Crayfish are very adept at capturing small fish with their chelae (pinchers), and can catch fish up to 2 inches in length. Those fish that are caught but escape can be damaged. Wounded areas on fish can develop bacterial or fungal infections that may eventually kill the fish. It is not necessary to feed crawfish in the pond unless you plan to raise large quantities for the home freezer or for sales.

Crayfish are generally classed as detritovores, scavengers or omnivores and feed low on the food chain. They recycle detrital material into animal flesh that is consumed by many fish species, turtles, frogs, wading birds, snakes alligators and raccoons. When available, crayfish are often the most important food item of the largemouth bass and are a very good fishing bait for this species and other fish.

Burrows

Burrows are dug into banks along the streams, ditches, ponds and other bodies of water inhabited by the crayfish. Burrows can also be found in marshy areas and wet pastures. Some species living in permanent waters do not build burrows, although they may scoop out a slanted depression in the substrate to rest in during the day.

Burrows are constructed in autumn or at any time when water level begins to fall and are built at night. Each burrow usually contains only one crayfish except during breeding season. However, young crayfish also may occasionally stay in a burrow inhabited by an adult.

Length and depth of the burrow depends on distance to the water table. Some burrows are over 3 feet deep. Burrows are dug horizontally into the bank or may be constructed at a sharp angle downward. Burrows end in a chamber where the crayfish spends most of the day. The chamber must be at least partially filled with water.

Chimneys at the mouth of burrow are created as a result of burrow construction. Soil dug by the crayfish is formed into a ball or oblong pellet and carried to the surface where it is deposited around the entrance. Material naturally accumulates and forms the chimney. The chimney has no known function however, it has been suggested that it may aid in ventilation of the burrow.

Harvesting Crayfish

Crayfish can be harvested in the traditional way by tying about ½ of a bacon strip or other fresh meat onto a length of string, casting it into shallow water that contains rocks or weed beds and waiting for the crayfish to grab the bait with its pinchers. When the crayfish "bites", slowly pull in the string and it will continue to hold the meat in the pinchers. When the creature is near enough to shore, scoop it with a small dipnet.

Funnel traps can also be used to catch crayfish. These traps are constructed of chicken wire, plastic mesh or hail screen. Plastic mesh or plastic coated welded wire have the greatest life span.

Form the mesh into a cylinder about 12 inches in diameter and 3 feet long. Construct a round mesh floor at one end of the cylinder. Attach the floor to the cylinder with wire. Two funnel shaped openings are placed at the bottom edge of the trap. The small end of the funnel is approximately 1 inch in diameter. The funnel extends into the trap about 4-6 inches and the funnel opening is about 2-3 inches in diameter. The top of the trap cylinder can be pressed flat and secured with wire or string to prevent crayfish escape but allow for the trap to be opened and crawfish harvested. Bait the trap with large dog biscuits, dead fish or any similar item. Place the trap in 1-4 feet of water near rocky areas or weed outcrops. Stake upright with a length of 3/8 inch diameter rebar or other suitable material. Traps should be harvested daily.

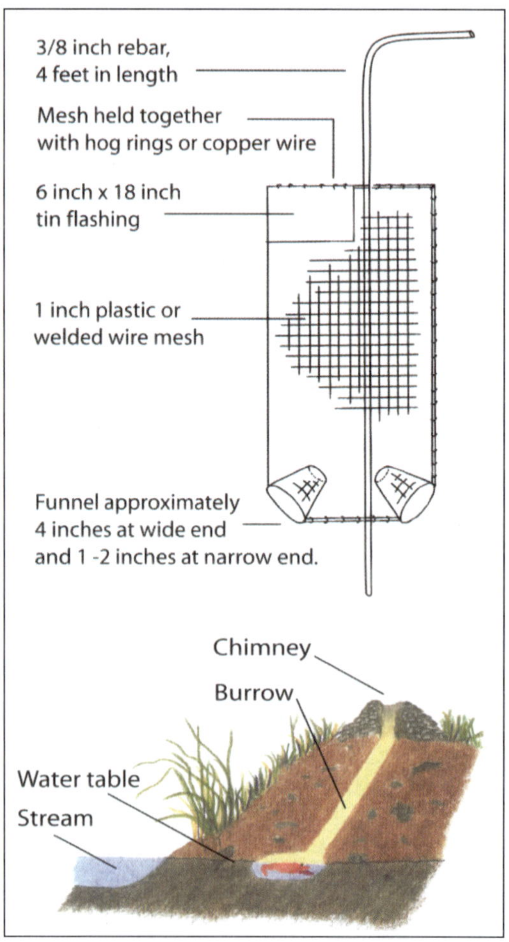

Typical crayfish burrow.

Handling Crayfish

Crayfish pinchers are quick and strong. In many species there is a spine at the tip of each pincher used to hold prey. The spines will easily pierce skin and may cause a little bleeding. Adult crayfish can give a nasty pinch to the unwary. The easiest way to handle these creatures is to net them with a small dip net. They can be picked up without pinching if they are firmly grasped by the thumb and index finger on both sides of the carapace.

Crayfish need well aerated water to survive and reproduce. In natural situations where dissolved oxygen values in water fall below 4-5 ppm, the crawfish will often climb out of the water to better aerate their gills and may begin traveling overland in search of a more favorable environment.

Commercial Production

Commercial production of crayfish in the U.S. is centered in Louisiana. It is often part of a field rotation with rice production. Crawfish are stocked at a rate of 20-100 lb/acre depending on the presence of existing stocks. Harvest varies with management intensity. Well managed open ponds can consistently produce 800-1500 lb/acre of crayfish annually. Rice ponds produce 300-500 lb/ acre /year while marsh ponds can produce up to 300 lb/acre/year of crayfish.

Water Mites- The Hydracarina

Hydracarina is not a taxonomic classification but a term of convenience for a group of over 600 species of similar, spider-like organisms. Water mites are 0.4 -3mm in diameter, about the size of a pinhead. They are often brilliantly colored, red, blue, green, yellow or a combination of these colors. Many of the mites are also mottled and brown in color.

The head, thorax and abdomen are fused into a singular body that is round, ovoid or globular in shape or sometimes flattened. The body may be soft or consist of leather-like plates.

Mites breathe through the skin. Many species have swimming hairs on the legs to assist with movement through the water. Most species have poor or little swimming ability and are active during daylight hours.

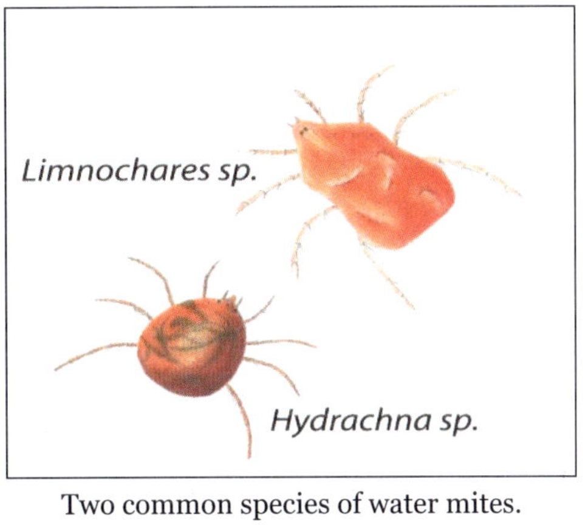

Two common species of water mites.

Water mites are found in streams, lakes, ponds and other kinds of aquatic habitat. They are most often found in littoral regions in densely vegetated water less than 2 m deep. Many species live on the bottom but are less likely to be seen than brightly colored species living in the vegetation or attached to aquatic insects.

Newly hatched larvae have 6 legs and spend a short quiet period in the water before parasitizing a host, usually an aquatic insect. The parasitic phase of the mite feeds off bodily fluids and may inject the host with digestive juices that liquify host tissues. Parasitized hosts seem unharmed by the mites.

Mites remain attached to insects such as dragonfly larvae even when they molt, become adults and fly away. After feeding on the host, the mite enters a brief resting period followed by metamorphosis into a nymph and falls off the host insect. It must land in the water to survive and reach sexually mature adulthood.

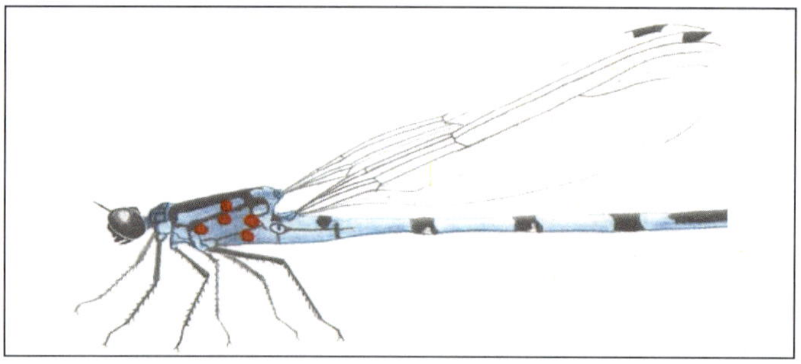

Mites attached to the body of a damsel fly.

Most nymphs mature in autumn. Eggs are deposited on stones or other bottom debris in groups of 20-400. Eggs are usually red and may be deposited singly or embedded in a gelatinous mass. Hatching takes place in 1-6 weeks. Mites probably live less than 1 year.

Adult mites feed on dead worms, small aquatic insects, ostracods, and other planktonic crustaceans. Mites are preyed on by carnivorous insects and hydra. They are not an important diet item of fish.

Mites can be collected by washing them off vegetation placed in an enamel pan or by stirring up pond bottom debris and straining the suspended material in a small, fine mesh dipnet.

Snails

Snails of many species are commonly found in ponds. Usually more snails can be found in heavily vegetated ponds than in bare, muddy ponds. Although large snail populations can be found in turbid aquaculture ponds where they feed on waste fish food rather than their natural diet of algae and aquatic plants.

Snails are a diet item for redear sunfish and freshwater drum. The redear sunfish, also known as the "shell cracker", is often stocked into ponds to control excessive snail populations. Snails of the genus helisoma are intermediate hosts of both the yellow and black grubs often found in sportfish taken from heavily vegetated ponds. Reducing snail populations and aquatic vegetation reduces grub infestation in sportfish.

Taxonomically, snails are divided into several large and complex groups. For the purposes of this article the main group of interest is the prosobranchs or Mesogastropoda which have a gill; and often, an operculum or door to close off the shell. Freshwater representatives of this group are found in rivers and lakes. Within the prosobranchs is a group called the Pulmonata, This group of snails have a sac-like lung and no operculum. These snails are by far the more common group in most ponds.

Pulmonate snails can breathe air at the surface through the rudimentary lung but can also survive indefinitely underwater by taking oxygen in through the skin. They prefer dissolved oxygen levels of at least 2 mg/l but can survive short periods of oxygen depletion by coming to the surface for air.

Snails lay eggs in gelatinous masses on the underside of leaves, sticks, stones and other debris found in the pond.

Snails inhabit and feed on aquatic vegetation and are usually found in shallow water but may be found at the lowest depths of aquatic vegetation growth. Snails scrape and shred plant material with a file-like tongue called a radula. The radula is covered with rows of microscopic teeth. Radula structure is one of the key identifiers currently used to classify snails into taxonomic groups.

Snails move by gliding on a film of slime secreted in their foot. Slime may function as a lubricant or is involved in some other way with locomotion. Muscles of the foot contract in waves that propel the creature forwards or backwards. Cilia, small hairs located on the bottom of the foot may also be involved in snail movement. Exact means of locomotion are unknown.

Lymaea sp., Physella sp. and Helisoma sp. Also move through the water by "spinning". The snail attaches slime to an object on the surface or on the substrate. It curls its foot laterally into a tube and emits a trail of slime as it moves toward the surface or bottom of the pond. The particular snail or other snails may use this slime track to move up and down through the water column until it becomes brittle and breaks apart.

Physella sp. Tadpole snail.

Three pulmonate families are most likely to be found in ponds, the Physidae, Lymnaeidae and Planorbidae.

The Physidae family, genus Physella sp. are by far the most common group of snails to be found in most ponds. The physella sp. Have many common names including: pond snails, pouch snails, tadpole snails and bladder snails. These snails can be up to 12 mm in length but are more often 6-8mm or less. Shell color is cream, through gray to black. The snail is most easily identified by the sinistral curve of the shell. If the shell is held facing you, the opening will be on the left hand side.

The Lymnaeidae are the largest of the pond snails. The genus Lymnaea is known as the great pond snail and can attain a length of 45- 60 mm. The shell opens dexterally or on the right side of the shell when the opening is viewed facing the observer. Lymnaea species a have long tapered spire and are colored yellow gray to black.

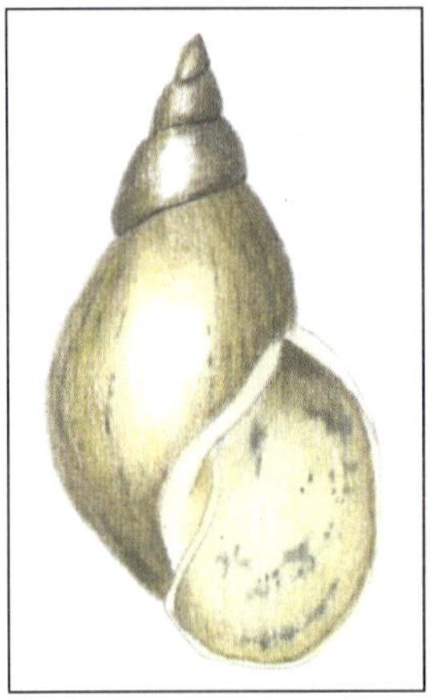

Lymnaea sp. Great pond snail.

Genus Helisoma sp. Is also known as the ram's horn snail. Other snails in the family Planorbidae also go by this common name. Helisoma sp. Are 11-15 mm in diameter. They are colored yellow-gray to black. This genus is parasitzed by larval stages of an avian tapeworm that is responsible for yellow grubs in fish. The tapeworm larvae burrow into the snail and feed on the digestive gland and other internal organs. Although much damage to the snail may result, the larvae do not often kill the snail and burrow out again after going through a further developmental stage.

Helisoma sp. Ram's horn snail.

Mussels

A variety of mussel species are to be found in North American freshwaters. Common native varieties include: lampsilis sp., Muckets, pinks, razorbacks and washboard, Corbicula fluminea the Asiatic or fingernail clam and the recently introduced exotic species, Dreissena polymorpha, the zebra mussel.

Freshwater mussels.

Mussels are bottom dwelling organisms that filter water to get their food. Water enters the mussel through the incurrent siphon and circulates over the gills and labial palps before exiting at the excurrent siphon. The labial palps direct particles of food such as phytoplankton and zooplankton into the mouth. Larger particles are passed out of the creature through the excurrent siphon.

The filtering action of mussels can have a profound effect on water clarity. Large mussel beds can remove much of the turbidity in water caused by plankton and suspended sediments. Resulting low levels of phytoplankton can reduce the productivity of the aquatic ecosystem because phytoplankton are a key component of the base of the aquatic food web.

Normally, mussels do not reach concentrations that adversely affect the environment, however, exotic species such as the zebra mussel, can out compete native species and reach population levels that impact the environment. In polluted aquatic environments where phytoplankton levels are very high, mussels can have a beneficial effect by filtering the water and removing many of these organisms.

Collecting Mussels

Mussels can be found in most aquatic environments. Large muckets can be found half to almost completely buried in silt or mud. They are easiest found when water levels in streams drop during mid to late summer. Freshly opened shells lying in the edge of the stream or on the banks indicate that raccoons or other animals may have been feeding on them recently. Look in nearby shallow water to find living specimens.

Their trails through mud are also visible during periods of low water flow. A sturdy garden rake can be a useful tool for finding mussels in mud or silt.

Look in a variety of aquatic habitats to increase the number of species be found. Some mussels prefer swift flowing water and a hard gravelly stream bottom. Others prefer slow water flow and a sand silt or mud substrate. Mussels will remain alive and can be transported in buckets without water for several hours if they are kept cool.

Corbicula fulminea, the asiatic or fingernail clam, can be found around the shores of many lakes. They congregate near, and in areas where water is flowing into the lake or pond. Large concentrations of these mussels can be found below outflow pipes or other areas where water is regularly discharged.

This mussel has commercial value in the "health food" industry. The entire mussel is boiled and the resulting broth is dried, powdered and made into capsules or tablets. It is sold as a health food supplement called "Corbicula" that is said to be good for your liver. Each teaspoon equals about 100 fresh mussels.

Marine Invertebrates

Ernst Haeckel's 96th plate, showing some marine invertebrates. Marine invertebrates have a large variety of body plans, which are currently categorised into over 30 phyla.

Marine invertebrates are the invertebrates that live in marine habitats. Invertebrate is a blanket term that includes all animals apart from the vertebrate members of the chordate phylum. Invertebrates lack a vertebral column, and some have evolved a shell or a hard exoskeleton. As on land and in the air, marine invertebrates have a large variety of body plans, and have been categorised into over 30 phyla. They make up most of the macroscopic life in the oceans.

Evolution

Dickinsonia may be the earliest animal.
They appear in the fossil record 571 million to 541 million years ago.

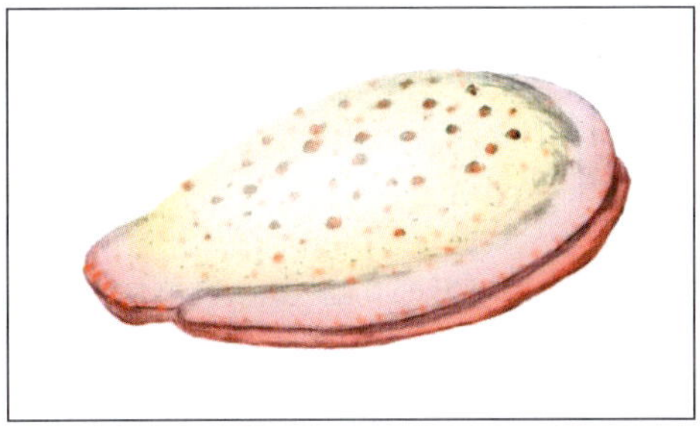

Kimberella, an early mollusc important for understanding the Cambrian explosion. Invertebrates are grouped into different phyla (body plans).

Opabinia, an extinct stem group arthropod appeared in the Middle Cambrian.

The earliest animals were marine invertebrates, that is, vertebrates came later. Animals are multicellular eukaryotes, and are distinguished from plants, algae, and fungi by lacking cell walls. Marine invertebrates are animals that inhabit a marine environment apart from the vertebrate members of the chordate phylum; invertebrates lack a vertebral column. Some have evolved a shell or a hard exoskeleton.

The earliest animals may belong to the genus *Dickinsonia*, 571 million to 541 million years ago. Individual *Dickinsonia* typically resemble a bilaterally symmetrical ribbed oval. They kept growing until they were covered with sediment or otherwise killed, and spent most of their lives with their bodies firmly anchored to the sediment. Their taxonomic affinities are presently unknown, but their mode of growth is consistent with a bilaterian affinity.

Apart from *Dickinsonia*, the earliest widely accepted animal fossils are the rather modern-looking cnidarians (the group that includes jellyfish, sea anemones and *Hydra*), possibly from around 580 Ma The Ediacara biota, which flourished for the last 40 million years before the start of the Cambrian, were the first animals more than a very few centimetres long. Like *Dickinsonia*, many were flat with a "quilted" appearance, and seemed so strange that there was a proposal to classify them as a separate kingdom, Vendozoa. Others, however, have been interpreted as early molluscs (*Kimberella*), echinoderms (*Arkarua*), and arthropods (*Spriggina*, *Parvancorina*). There is still debate about the classification of these specimens, mainly because the diagnostic features which allow taxonomists to classify more recent organisms, such as similarities to living organisms, are generally absent in the Ediacarans. However, there seems little doubt that *Kimberella* was at least a triploblastic bilaterian animal, in other words, an animal significantly more complex than the cnidarians.

The small shelly fauna are a very mixed collection of fossils found between the Late Ediacaran and Middle Cambrian periods. The earliest, *Cloudina*, shows signs of successful defense against predation and may indicate the start of an evolutionary arms race. Some tiny Early Cambrian shells almost certainly belonged to molluscs, while the owners of some "armor plates," *Halkieria* and *Microdictyon*, were eventually identified when more complete specimens were found in Cambrian lagerstätten that preserved soft-bodied animals.

In the 1970s there was already a debate about whether the emergence of the modern phyla was "explosive" or gradual but hidden by the shortage of Precambrian animal fossils. A re-analysis of fossils from the Burgess Shale lagerstätte increased interest in the issue when it revealed animals, such as *Opabinia*, which did not fit into any known phylum. At the time these were interpreted as evidence that the modern phyla had evolved very rapidly in the Cambrian explosion and that the Burgess Shale's "weird wonders" showed that the Early Cambrian was a uniquely experimental period of animal evolution. Later discoveries of similar animals and the development of new theoretical approaches led to the conclusion that many of the "weird wonders" were evolutionary "aunts" or "cousins" of modern groups—for example that *Opabinia* was a member of the lobopods, a group which includes the ancestors of the arthropods, and that it may have been closely related to the modern tardigrades. Nevertheless, there is still much debate about whether the Cambrian explosion was really explosive and, if so, how and why it happened and why it appears unique in the history of animals.

Classification

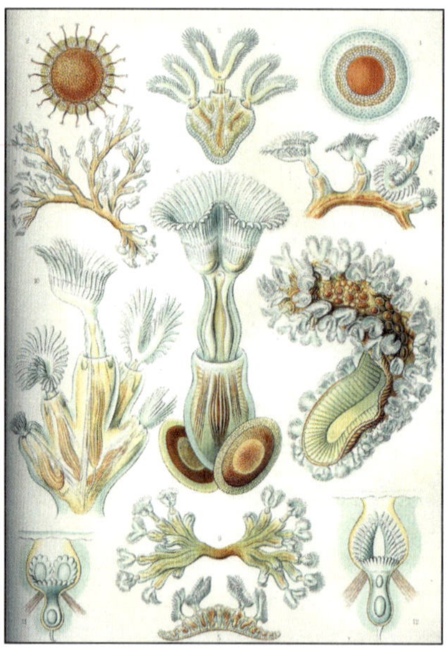

Bryozoa, from Ernst Haeckel's *Kunstformen der Natur*, 1904.

Invertebrates are grouped into different phyla. Informally phyla can be thought of as a way of grouping organisms according to their body plan. A body plan refers to a blueprint which describes the shape or morphology of an organism, such as its symmetry, segmentation and the disposition of its appendages. The idea of body plans originated with vertebrates, which were grouped into one phylum. But the vertebrate body plan is only one of many, and invertebrates consist of many phyla or body plans. The history of the discovery of body plans can be seen as a movement from a worldview centred on vertebrates, to seeing the vertebrates as one body plan among many. Among the pioneering zoologists, Linnaeus identified two body plans outside the vertebrates; Cuvier identified three; and Haeckel had four, as well as the Protista with eight more, for a total of twelve. For comparison, the number of phyla recognised by modern zoologists has risen to 35.

Historically body plans were thought of as having evolved in rapidly during the Cambrian explosion, but a more nuanced understanding of animal evolution suggests a gradual development of body plans throughout the early Palaeozoic and beyond. More generally a phylum can be defined in two ways: as described above, as a group of organisms with a certain degree of morphological or developmental similarity (the phenetic definition), or a group of organisms with a certain degree of evolutionary relatedness (the phylogenetic definition).

As on land and in the air, invertebrates make up a great majority of all macroscopic life, as the vertebrates makes up a subphylum of one of over 30 known animal phyla, making the term almost meaningless for taxonomic purpose. Invertebrate sea life includes the following groups, some of which are phyla:

Various sea anemones classified as Actiniae, in the Cnidaria phylum.

"A variety of marine worms".

- Acoela, among the most primitive bilateral animals;
- Annelida, (polychaetes and sea leeches);
- Brachiopoda, marine animals that have hard "valves" (shells) on the upper and lower surfaces;
- Bryozoa, also known as moss animals or sea mats;
- Chaetognatha, commonly known as arrow worms, are a phylum of predatory marine worms that are a major component of plankton;
- Cephalochordata represented in the modern oceans by the lancelets (also known as Amphioxus);
- Cnidaria, such as jellyfish, sea anemones, and corals;
- Crustacea, including lobsters, crabs, shrimp, crayfish, barnacles, hermit crabs, mantis shrimps, and copepods;

- Ctenophora, also known as comb jellies, the largest animals that swim by means of cilia;
- Echinodermata, including sea stars, brittle stars, sea urchins, sand dollars, sea cucumbers, crinoids, and sea daisies;
- Echiura, also known as spoon worms;
- Gnathostomulids, slender to thread-like worms, with a transparent body that inhabit sand and mud beneath shallow coastal waters;
- Gastrotricha, often called hairy backs, found mostly interstitially in between sediment particles;
- Hemichordata, includes acorn worms, solitary worm-shaped organisms;
- Kamptozoa, goblet-shaped sessile aquatic animals, with relatively long stalks and a "crown" of solid tentacles, also called Entoprocta;
- Kinorhyncha, segmented, limbless animals, widespread in mud or sand at all depths, also called mud dragons;
- Loricifera, very small to microscopic marine sediment-dwelling animals only discovered in 1983;
- Mollusca, including shellfish, squid, octopus, whelks, Nautilus, cuttlefish, nudibranchs, scallops, sea snails, Aplacophora, Caudofoveata, Monoplacophora, Polyplacophora, and Scaphopoda;
- Myzostomida, a taxonomic group of small marine worms which are parasitic on crinoids or "sea lilies";
- Nemertinea, also known as "ribbon worms" or "proboscis worms";
- Orthonectida, a small phylum of poorly known parasites of marine invertebrates that are among the simplest of multi-cellular organisms;
- Phoronida, a phylum of marine animals that filter-feed with a lophophore (a "crown" of tentacles), and build upright tubes of chitin to support and protect their soft bodies;
- Placozoa, small, flattened, multicellular animals around 1 millimetre across and the simplest in structure. They have no regular outline, although the lower surface is somewhat concave, and the upper surface is always flattened;
- Porifera (sponges), multicellular organisms that have bodies full of pores and channels allowing water to circulate through them;
- Priapulida, or penis worms, are a phylum of marine worms that live marine mud. They are named for their extensible spiny proboscis, which, in some species, may have a shape like that of a human penis;
- Pycnogonida, also called sea spiders, are unrelated to spiders, or even to arachnids which they resemble;

- Sipunculida, also called peanut worms, is a group containing 144–320 species (estimates vary) of bilaterally symmetrical, unsegmented marine worms;
- Tunicata, also known as sea squirts or sea pork, are filter feeders attached to rocks or similarly suitable surfaces on the ocean floor;
- Some flatworms of the classes Turbellaria and Monogenea;
- Xenoturbella, a genus of bilaterian animals that contains only two marine worm-like species.
- Xiphosura, includes a large number of extinct lineages and only four recent species in the family Limulidae, which include the horseshoe crabs.

Arthropods total about 1,113,000 described extant species, molluscs about 85,000 and chordates about 52,000.

Marine Sponges

Sponges are animals of the phylum Porifera (Modern Latin for *bearing pores*). They are multicellular organisms that have bodies full of pores and channels allowing water to circulate through them, consisting of jelly-like mesohyl sandwiched between two thin layers of cells. They have unspecialized cells that can transform into other types and that often migrate between the main cell layers and the mesohyl in the process. Sponges do not have nervous, digestive or circulatory systems. Instead, most rely on maintaining a constant water flow through their bodies to obtain food and oxygen and to remove wastes.

Sponges are similar to other animals in that they are multicellular, heterotrophic, lack cell walls and produce sperm cells. Unlike other animals, they lack true tissues and organs, and have no body symmetry. The shapes of their bodies are adapted for maximal efficiency of water flow through the central cavity, where it deposits nutrients, and leaves through a hole called the osculum. Many sponges have internal skeletons of spongin and/or spicules of calcium carbonate or silicon dioxide. All sponges are sessile aquatic animals. Although there are freshwater species, the great majority are marine (salt water) species, ranging from tidal zones to depths exceeding 8,800 m (5.5 mi).

Sponges have no nervous, digestive or circulatory system.

While most of the approximately 5,000–10,000 known species feed on bacteria and other food particles in the water, some host photosynthesizing micro-organisms as endosymbionts and these alliances often produce more food and oxygen than they consume. A few species of sponge that live in food-poor environments have become carnivores that prey mainly on small crustaceans.

Sponge biodiversity. There are four sponge species in this photo.

Venus' flower basket at a depth of 2572 meters.

Barrel sponge.

Linnaeus mistakenly identified sponges as plants in the order Algae. For a long time thereafter sponges were assigned to a separate subkingdom, Parazoa (meaning *beside the animals*). They are now classified as a paraphyletic phylum from which the higher animals have evolved.

Marine Cnidarians

Cnidarians are the simplest animals with cells organised into tissues.
Yet the starlet sea anemone contains the same genes as those that form the vertebrate head.

Cnidarians are distinguished by the presence of stinging cells, specialized cells that they use mainly for capturing prey. Cnidarians include corals, sea anemones, jellyfish and hydrozoans. They form a phylum containing over 10,000 species of animals found exclusively in aquatic (mainly marine) environments. Their bodies consist of mesoglea, a non-living jelly-like substance, sandwiched between two layers of epithelium that are mostly one cell thick. They have two basic body forms: swimming medusae and sessile polyps, both of which are radially symmetrical with mouths surrounded by tentacles that bear cnidocytes. Both forms have a single orifice and body cavity that are used for digestion and respiration.

Fossil cnidarians have been found in rocks formed about 580 million years ago. Fossils of cnidarians that do not build mineralized structures are rare. Scientists currently think cnidarians, ctenophores and bilaterians are more closely related to calcareous sponges than these are to other sponges, and that anthozoans are the evolutionary "aunts" or "sisters" of other cnidarians, and the most closely related to bilaterians.

Cnidarians are the simplest animals in which the cells are organised into tissues. The starlet sea anemone is used as a model organism in research. It is easy to care for in the laboratory and a protocol has been developed which can yield large numbers of embryos on a daily basis. There is a remarkable degree of similarity in the gene sequence conservation and complexity between the sea anemone and vertebrates. In particular, genes concerned in the formation of the head in vertebrates are also present in the anemone.

Sea anemones are common in tidepools.

Their tentacles sting and paralyse small fish.

Close up of polyps on the surface of a coral, waving their tentacles.

Marine Worms

Arrow worms are predatory components of plankton worldwide.

Worms (Old English for *serpent*) typically have long cylindrical tube-like bodies and no limbs. Marine worms vary in size from microscopic to over 1 metre (3.3 ft) in length for some marine polychaete worms (bristle worms) and up to 58 metres (190 ft) for the marine nemertean worm (bootlace worm). Some marine worms occupy a small variety of parasitic niches, living inside the bodies of other animals, while others live more freely in the marine environment or by burrowing underground.

Different groups of marine worms are related only distantly, so they are found in several different phyla such as the Annelida (segmented worms), Chaetognatha (arrow worms), Hemichordata, and Phoronida (horseshoe worms). Many of these worms have specialized tentacles used for exchanging oxygen and carbon dioxide and also may be used for reproduction. Some marine worms are tube worms, such as the giant tube worm which lives in waters near underwater volcanoes and can withstand temperatures up to 90 degrees Celsius.

Platyhelminthes (flatworms) form another worm phylum which includes a class Cestoda of parasitic tapeworms. The marine tapeworm *Polygonoporus giganticus*, found in the gut of sperm whales, can grow to over 30 m (100 ft).

Giant tube worms cluster around hydrothermal vents.

Nematodes (roundworms) constitute a further worm phylum with tubular digestive systems and an opening at both ends. Over 25,000 nematode species have been described, of which more than half are parasitic. It has been estimated another million remain undescribed. They are ubiquitous in marine, freshwater and terrestrial environments, where they often outnumber other animals in both individual and species counts. They are found in every part of the earth's lithosphere, from the top of mountains to the bottom of oceanic trenches. By count they represent 90% of all animals on the ocean floor. Their numerical dominance, often exceeding a million individuals per square meter and accounting for about 80% of all individual animals on earth, their diversity of life cycles, and their presence at various trophic levels point at an important role in many ecosystems.

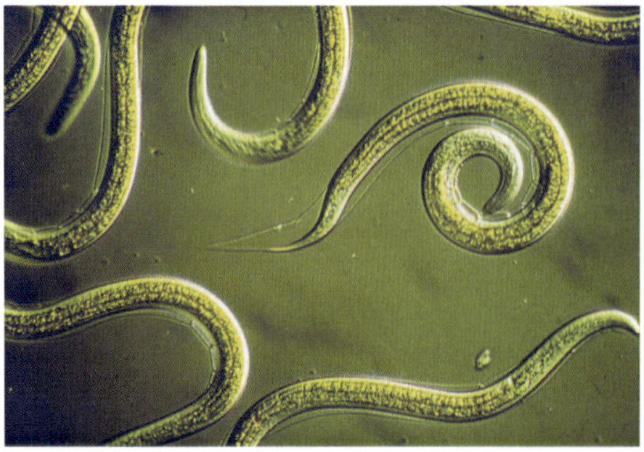

Nematodes are ubiquitous pseudocoelomates which can parasite marine plants and animals.

Bloodworms are typically found on the bottom of shallow marine waters.

Bobbit worms are ambush predators that live on the seafloor.

Echinoderms

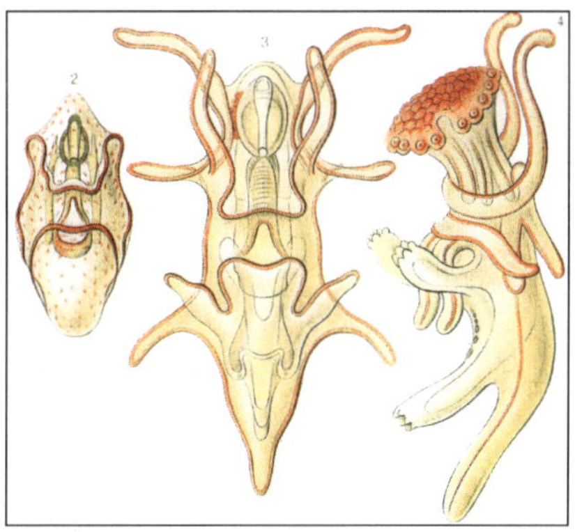

Starfish larvae are bilaterally symmetric, whereas the adults have fivefold symmetry.

Echinoderms is a phylum which contains only marine invertebrates. The adults are recognizable by their radial symmetry (usually five-point) and include starfish, sea urchins, sand dollars, and sea cucumbers, as well as the sea lilies. Echinoderms are found at every ocean depth, from the intertidal zone to the abyssal zone. The phylum contains about 7000 living species, making it the second-largest grouping of deuterostomes (a superphylum), after the chordates (which include the vertebrates, such as birds, fishes, mammals, and reptiles).

Echinoderms are unique among animals in having bilateral symmetry at the larval stage, but fivefold symmetry (pentamerism, a special type of radial symmetry) as adults.

The echinoderms are important both biologically and geologically. Biologically, there are few other groupings so abundant in the biotic desert of the deep sea, as well as shallower oceans. Most echinoderms are able to regenerate tissue, organs, limbs, and reproduce asexually; in some cases, they can undergo complete regeneration from a single limb. Geologically, the value of echinoderms is in their ossified skeletons, which are major contributors to many limestone formations, and can provide valuable clues as to the geological environment. They were the most used species in regenerative research in the 19th and 20th centuries. Further, it is held by some scientists that the radiation of echinoderms was responsible for the Mesozoic Marine Revolution.

Echinoderm literally means "spiny skin", as this water melon sea urchin illustrates.

Sea cucumbers filter feed on plankton and suspended solids.

The ochre sea star was the first keystone predator to be studied. They limit mussels which can overwhelm intertidal communities.

Aside from the hard-to-classify *Arkarua* (a Precambrian animal with echinoderm-like pentamerous radial symmetry), the first definitive members of the phylum appeared near the start of the Cambrian.

Marine Molluscs

Reconstruction of an ammonite, a highly successful early cephalopod that first appeared in the Devonian (about 400 mya). They became extinct during the same extinction event that killed the land dinosaurs (about 66 mya).

Molluscs form a phylum with about 85,000 extant recognized species. By species count they are the largest marine phylum, comprising about 23% of all the named marine organisms. Molluscs have more varied forms than other invertebrate phylums. They are highly diverse, not just in size and in anatomical structure, but also in behavior and in habitat. The majority of species still live in the oceans, from the seashores to the abyssal zone, but some form a significant part of the freshwater fauna and the terrestrial ecosystems.

The mollusc phylum is divided into 9 or 10 taxonomic classes, two of which are extinct. These classes include gastropods, bivalves and cephalopods, as well as other lesser-known but distinctive classes. Gastropods with protective shells are referred to as snails (sea snails), whereas gastropods without protective shells are referred to as slugs (sea slugs). Gastropods are by far the most numerous molluscs in terms of classified species, accounting for 80% of the total. Bivalves include clams, oysters, cockles, mussels, scallops, and numerous other families. There are about 8,000 marine bivalves species (including brackish water and estuarine species), and about 1,200 freshwater species. Cephalopod include octopus, squid and cuttlefish. They are found in all oceans, and neurologically are the most advanced of the invertebrates. About 800 living species of marine cephalopods have been identified, and an estimated 11,000 extinct taxa have been described. There are no fully freshwater cephalopods.

The nautilus is a living fossil little changed since it evolved 500 million years ago as one of the first cephalopods.

Marine gastropods are sea snails or sea slugs. This nudibranch is a sea slug.

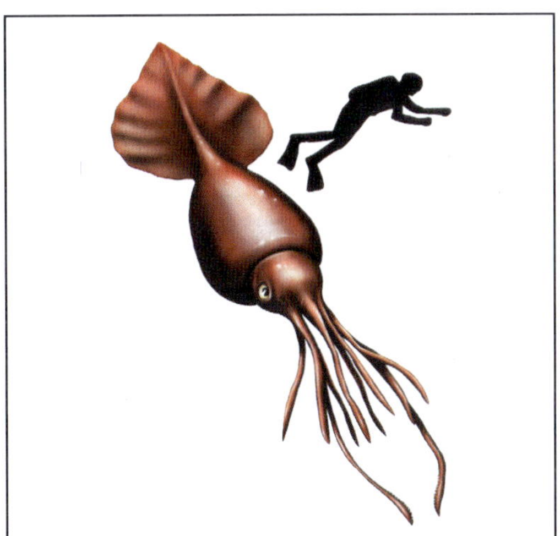

Colossal squid, largest of all invertebrates.

The sea snail *Syrinx aruanus* has the largest shell of any living gastropod.

Molluscs usually have eyes. Bordering the edge of the mantle of a scallop, a bivalve mollusc, can be over 100 simple eyes.

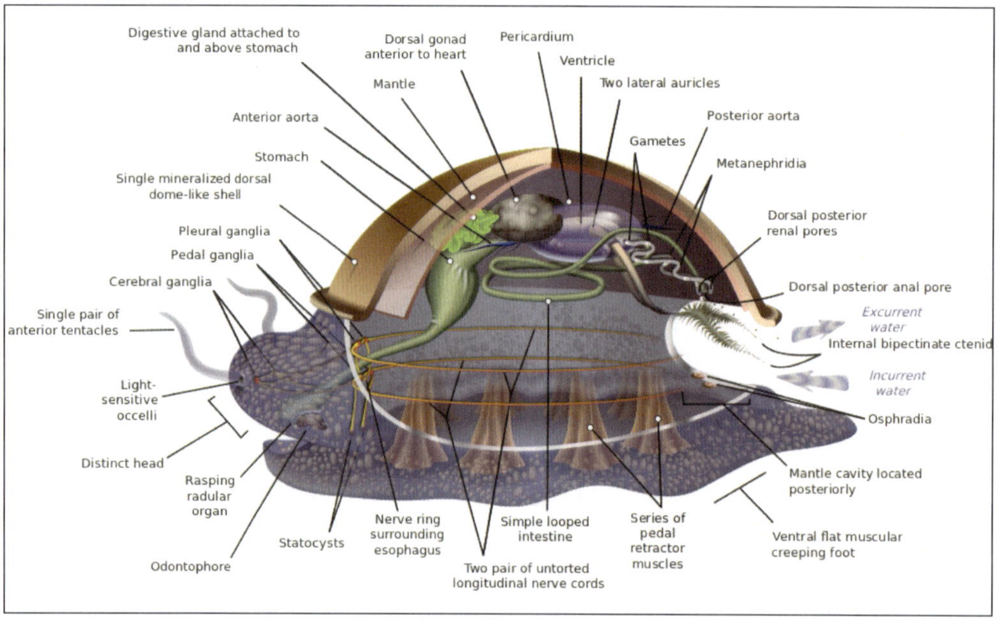

Generalized or *hypothetical ancestral mollusc*.

This generalized mollusc is unsegmented and bilaterally symmetrical with an underside consisting of a single muscular foot. Beyond that it has three further key features. Firstly, it has a muscular cloak called a mantle covering its viscera and containing a significant cavity used for breathing and excretion. A shell secreted by the mantle covers the upper surface. Secondly (apart from bivalves) it has a rasping tongue called a radula used for feeding. Thirdly, it has a nervous system including a complex digestive system using microscopic, muscle-powered hairs called cilia to exude mucus. The generalized mollusc has two paired nerve cords (three in bivalves). The brain, in species that have one, encircles the esophagus. Most molluscs have eyes and all have sensors detecting chemicals, vibrations, and touch. The simplest type of molluscan reproductive system relies on external fertilization, but more complex variations occur. All produce eggs, from which may emerge trochophore larvae, more complex veliger larvae, or miniature adults. The depiction is rather similar to modern monoplacophorans, and some suggest it may resemble very early molluscs.

Good evidence exists for the appearance of marine gastropods, cephalopods and bivalves in the Cambrian period 541 to 485.4 million years ago. However, the evolutionary history both of molluscs' emergence from the ancestral Lophotrochozoa and of their diversification into the well-known living and fossil forms are still subjects of vigorous debate among scientists.

Marine Arthropods

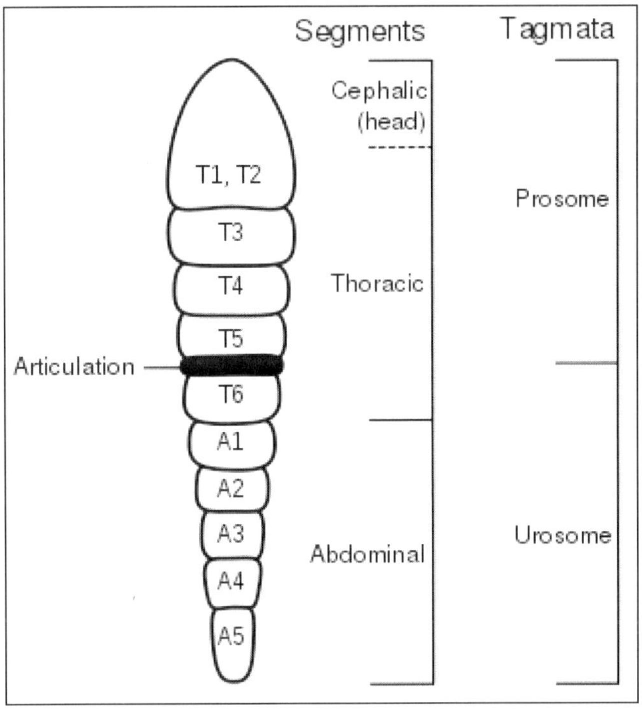

Segmentation and tagmata of an arthropod.

Arthropods have an exoskeleton (external skeleton), a segmented body, and jointed appendages (paired appendages). They form a phylum which includes insects, arachnids, myriapods, and crustaceans. Arthropods are characterized by their jointed limbs and cuticle made of chitin, often mineralised with calcium carbonate. The arthropod body plan consists of segments, each with a pair of appendages. The rigid cuticle inhibits growth, so arthropods replace it periodically by moulting. Their versatility has enabled them to become the most species-rich members of all ecological guilds in most environments.

Marine arthropods range in size from the microscopic crustacean *Stygotantulus* to the Japanese spider crab. Arthropods' primary internal cavity is a hemocoel, which accommodates their internal organs, and through which their haemolymph - analogue of blood - circulates; they have open circulatory systems. Like their exteriors, the internal organs of arthropods are generally built of repeated segments. Their nervous system is "ladder-like", with paired ventral nerve cords running through all segments and forming paired ganglia in each segment. Their heads are formed by fusion of varying numbers of segments, and their brains are formed by fusion of the ganglia of these segments and encircle the esophagus. The respiratory and excretory systems of arthropods vary, depending as much on their environment as on the subphylum to which they belong.

Their vision relies on various combinations of compound eyes and pigment-pit ocelli: in most species the ocelli can only detect the direction from which light is coming, and the compound eyes are the main source of information, but the main eyes of spiders are ocelli that can form images and, in a few cases, can swivel to track prey. Arthropods also have a wide range of chemical and mechanical sensors, mostly based on modifications of the many setae (bristles) that project through their cuticles. Arthropods' methods of reproduction and development are diverse; all terrestrial species use internal fertilization, but this is often by indirect transfer of the sperm via an appendage or the ground, rather than by direct injection. Marine species all lay eggs and use either internal or external fertilization. Arthropod hatchlings vary from miniature adults to grubs that lack jointed limbs and eventually undergo a total metamorphosis to produce the adult form.

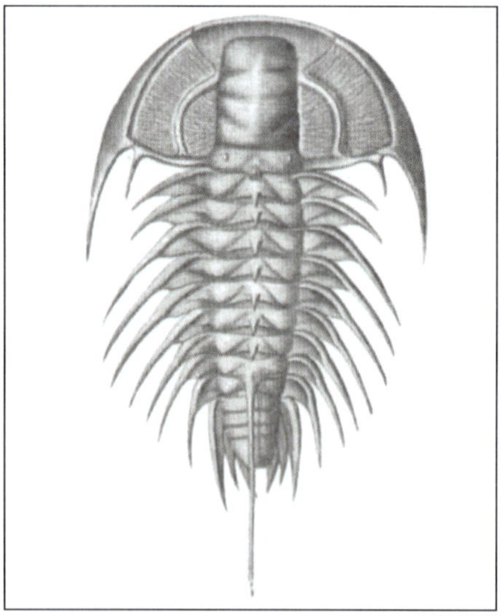

Trilobites, now extinct, roamed oceans for 270 million years.

Horseshoe crab, a living fossil arthropod from 450 million years ago Crustaceans

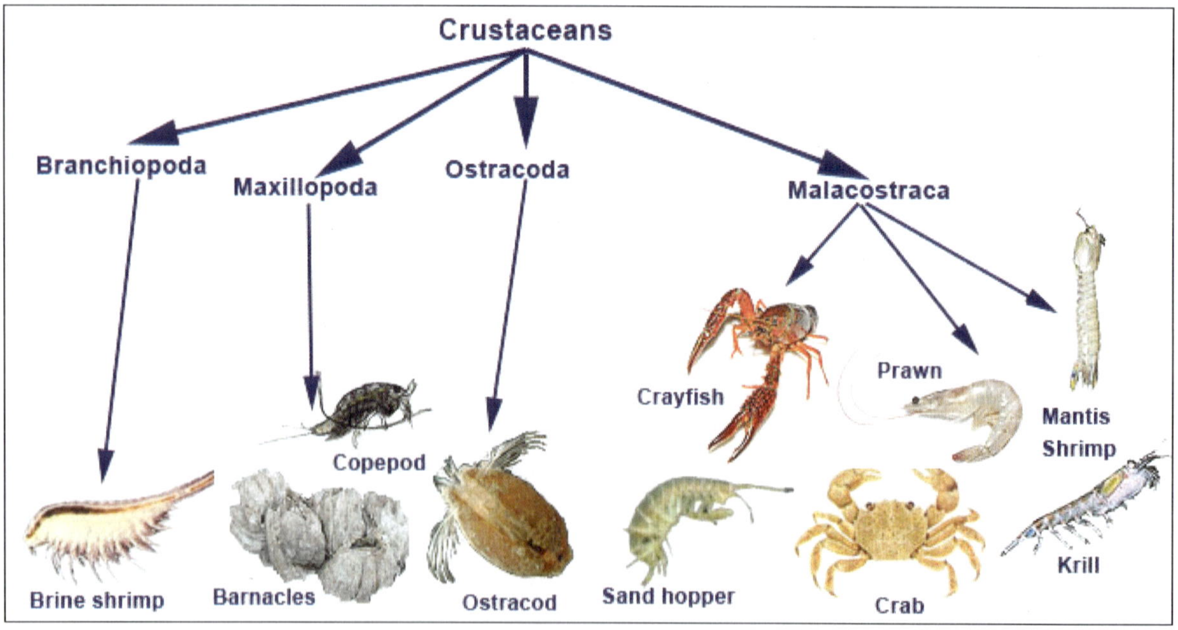

The evolutionary ancestry of arthropods dates back to the Cambrian period. The group is generally regarded as monophyletic, and many analyses support the placement of arthropods with cycloneuralians (or their constituent clades) in a superphylum Ecdysozoa. Overall however, the basal relationships of Metazoa are not yet well resolved. Likewise, the relationships between various arthropod groups are still actively debated.

Other Phyla

- Tardigrade, Lobopodia, (Onychophora)
- Non-craniate (non-vertebrate) chordates: Cephalochordate, Tunicata and Haikouella. These invertebrates are close relatives of the vertebrates.

Non-craniate Chordates are Close Relatives of Vertebrates

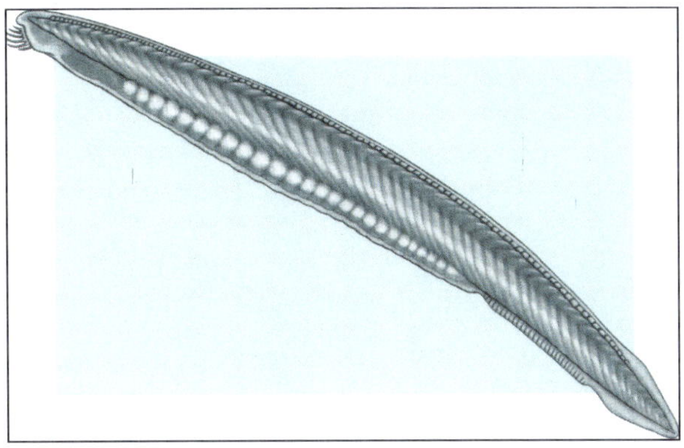

The lancelet, a small translucent fish-like Cephalochordate, is the closest living invertebrate relative of the vertebrates.

Fluorescent-colored sea squirts, *Rhopalaea crassa*.
Tunicates may provide clues to vertebrate (and therefore human) ancestry.

Salp chain.

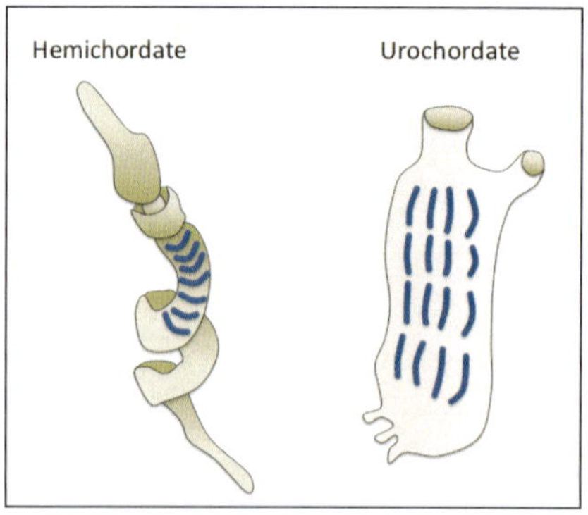

Gill slits in an acorn worm (left) and tunicate (right).

Minerals from Sea Water

There are a number of marine invertebrates that use minerals that are present in the sea in such minute quantities that they were undetectable until the advent of spectroscopy. Vanadium is concentrated by some tunicates for use in their blood cells to a level ten million times that of the surrounding seawater. Other tunicates similarly concentrate niobium and tantalum. Lobsters use copper in their respiratory pigment hemocyanin, despite the proportion of this metal in seawater being minute. Although these elements are present in vast quantities in the ocean, their extraction by man is not economic.

Terrestrial Invertebrates

Invertebrates affect all other forms of life by sheer weight of numbers. They make up the bulk of biodiversity and are essential for maintaining the function of ecosystems in many ways. For example, they form an integral part of the food web, they recycle organic matter, feeding on faeces or dead plants and animals, and they make up the bulk of parasitic species (along with microorganisms such as bacteria) which regulate animal and plant numbers.

Almost all species from the region have much broader distributions and can usually be found in adjacent regions of New South Wales and Victoria. Given that the number and diversity of species in the region is enormous, exemplar taxa are used to indicate the breadth and diversity of terrestrial invertebrates in the region.

Platyhelminthes (Flat Worms)

Platyhelminthes are soft-bodied, generally flattened, worm-like, unsegmented animals, with a wide range of body forms. The phylum is divided into three classes: Turbellaria, Trematoda and Cestoda. The majority of aquatic flatworms are turbellarians. Trematodes (flukes) and cestodes (tapeworms) are parasitic although some trematodes have free-living dispersive stages. The Australian terrestrial cestode fauna is poorly documented, with the best-studied host groups being the marsupials. Overall, cestodes have been reported from only 5% of vertebrate species. Endemism among Australian flatworms is likely to be low in birds and high in parasites of frogs, reptiles and marsupials, and in free-living freshwater forms.

Quite a few trematode and cestode parasites of domestic animals occur in the region, including liver fluke of sheep and cattle (Fasciola hepatica), sheep whipworm (Trichuris ovis), sheep tapeworm (Moniezia expansa), and stomach hair worm (Trichostrongylus axei), which occurs commonly in sheep, other ruminants, and horses. Other examples include the tapeworm Echinococcus granulosus (hydatids), in which humans can act as intermediate hosts. Definitive hosts of this tapeworm are carnivores such as dogs, foxes and dingoes, while intermediate hosts are usually herbivores such as sheep, cattle, kangaroos, wallabies, and wombats. Cestodes are also commonly found in reptiles, birds, and marsupials in the region, and just one of many examples is Oochoristica trachysauri, a tapeworm from the sleepy lizard, Trachysaurus rugosus.

Nematodes (Round Worms)

Nematodes are unsegmented roundworms, which are not related to earthworms. They are commonly found in soil and freshwater habitats. Others are parasitic in roots and aerial parts of plants, in vertebrates, and in invertebrates such as insects. Free-living forms feed on living organisms such as bacteria rather than on dead organic matter. Discussion here is limited to terrestrial nematodes living in soil and/or parasitising plants. These nematodes are usually microscopic (most less than 3 mm long). Almost all soils contain nematodes, and in Riverland soils with high organic content there may be hundreds per gram of soil. In a survey of soil microinvertebrate assemblages, Bird found that nematodes dominated both in numbers and diversity in soils from the shores of Lake Alexandrina and the Murray River estuary at Goolwa (87 and 93% of total numbers of microinvertebrates, respectively). There is little knowledge of most Australian soil nematodes, and most species remain undescribed.

Plant parasites are the best known group of soil nematodes, due to their economic importance. These nematodes feed on plant roots, and occasionally within plant leaves. Tylenchids and aphelenchids have a hollow stylet or spear within the head, which they can insert into plant or fungal cells and through which they can suck out some of the cell contents. Other soil nematodes feed on bacteria, algae, fungal hyphae and other nematodes, and invertebrates such as protozoa, rotifers and enchytraeids. Rhabditid nematodes usually feed on bacteria, but some are predatory. They have a tube-like mouth, which may have teeth at its base. Rhabditids are important for the recycling of soil nutrients. Mononchids are predators, with teeth that they use to seize their prey. Dorylaimids include a large group of soil and freshwater nematodes, some of which feed on plants, others on bacteria and algae and possibly fungal hyphae, while some are predatory. The size and form of the spear is variable in dorylaimids; with long and slender spears in plant parasites, and short, hollow forms with wide openings in bacterial and algal feeders.

Soil nematodes are generally able to survive long periods of desiccation, important for their survival in many Riverland soils. They may survive dry periods as eggs, or they may coil and become inactive as the soil dries out, entering a state of anhydrobiosis in which metabolism is suspended. Whilst in this state, nematodes are brittle and their cuticle (body covering) is easily damaged by sharp soil particles. However, provided they are not damaged, they recover their normal functions when soil moisture rises again.

Many species of nematodes and a few genera associated with native plants are endemic to Australia, and several species are known only from South Australia. The greatest diversity of species is usually found in areas with moist soils, such as gullies. Only one study of nematodes from soils in the Riverland and Murraylands has been carried out (at Ferries MacDonald Conservation Park at Monarto; Frances Reay, unpublished data). Considering the similarity in climate, soils and vegetation, it is probable that the nematode fauna of the region is similar to that of adjoining regions in Victoria and New South Wales, where some survey work has been done. In broad scale farming, there are many records of economically important tylenchids such as cereal cyst nematode (Heterodera avenae) and root lesion nematode (Pratylenchus neglectus). Root knot nematodes (Meloidogyne spp.), citrus nematode (Tylenchulus semipenetrans) and the dorylaimid Xiphinema sp. are commonly found in horticultural soils. Sauertylenchus sp., Morulaimus whitei, and Paralongidorus eucalypti, all widespread in Australia, have been recorded from Ferries MacDonald Conservation Park. Also recorded were various rhabditids and dorylaimids, including stubby root nematode (Paratrichodorus sp.), which has a long stylet and feeds from outside a root, while the tylenchid genera Radopholus (burrowing nematodes), Scutellonema (spiral nematodes), Hemicriconemoides, Criconemoides (ring nematodes), and Hemicycliophora (sheath nematodes) are also known from this locality (Waite Insect and Nematode Collection, The University of Adelaide).

Shoot bud gall of Fergusonina lockharti.

The river red gum (Eucalyptus camaldulensis) is host to at least four different combinations of cecidogenic (gall-forming) species of the nematode Fergusobia and fly Fergusonina. These nematodes and flies together form unique, species-specific mutualisms, in which the nematode relies on the fly for transport, and the fly on the nematode for gall induction. Galls vary in form from large shoot bud galls, which support hundreds of flies and nematodes, to flower bud galls, 'stem' galls and small axial bud galls that only support a few flies and accompanying nematodes.

Annelida (Segmented Worms: Earthworms and Leeches)

Introduced earthworms (mostly Lumbrucidae), including Aporrectodea spp., are more likely to be encountered than native species, especially in disturbed areas such as pastures, orchards, vineyards and gardens. This was the case in a survey of citrus orchards between Waikerie, Loxton and Paringa. Lumbricids were introduced into Australia by European settlers, and are native to Britain and other European temperate regions. Once introduced into a new area, they have the ability to breed rapidly and colonise new areas, often surpassing the original native species. The tiger worm, red worm or composting worm (Eisenia fetida), which is widely sold for composting, also occurs in the region, although it prefers rotting vegetation, compost, and manure rather than soil.

Megascolecidae is the only indigenous family of earthworms in South Australia, and although the fauna is impoverished, species endemicity is relatively high. Jamieson suggests that the paucity of the fauna is correlated with the low rainfall over most of the state. Unfortunately, most research in South Australia on earthworms appears to have taken place in higher rainfall areas (e.g. Mt Lofty Ranges, Fleurieu Peninsula and the South-East). Species from the Riverland and Murraylands include Heteroporodrilus shephardi, collected near Bordertown and extending to the Wimmera River in Victoria, and Gemascolex lateralis, collected near Tailem Bend but more widely distributed in the Mt Lofty Ranges and Fleurieu Peninsula.

Mollusca (Snails and Slugs)

Introduced land snails are very common in the region and are often a severe problem. These include the common white snail or vineyard snail, white Italian snail (Theba pisana), conical/pointed snails, and small brown snail (Microxeromagna vestita). The garden snail (Helix aspersa) is also common in areas that are regularly irrigated (e.g. orchards and gardens). Cernuella virgata, T. pisana and Cochicella spp. can be seen aestivating in the drier months on trees, fence posts, buildings or vegetation. Aestivation is a form of dormancy where the snails seal off their aperture (opening) with a membrane and remain inactive for long periods of time.

(left), Common white snail, Cernuella virgata, aestivating on an old headstone at Black Hill Old Cemetery SA (Photograph: Gary Taylor); 3 (right), conical snail, Cochicella sp.

Two introduced slug species, Lehmannia flava and Milax gagates, are common in irrigated situations. The former species is larger and yellowish, whereas the latter species is smaller and black.

Crustaceans (Slaters and Land-hoppers)

Terrestrial crustaceans such as isopods (Isopoda) (slaters) from the region are poorly known, but include the introduced European woodlouse (Porcellio laevis), the common woodlouse or pillbug (Armadillidium vulgare) and the delicate slater (Porcellionides pruinosus). Slaters are commonly found under debris and leaf litter in moist situations. Land-hoppers (Amphipoda) such as Austrotriodes crenatus are also common in shadehouses.

Myriapods (Millipedes and Centipedes)

Millipedes (Diplopoda) are long, thin arthropods with two pairs of legs on most posterior body segments, which feed largely on decaying plant material. Native millipedes have been poorly studied in the region, however the familiar introduced black Portuguese millipede (Ommatoiulus moreletii) occurs throughout the region and is mostly seen during the wetter times of the year.

Centipedes (Chilopoda) are long, thin arthropods, which unlike millipedes, never have more than one pair of legs per body segment. They live in moist situations such as in rock crevices, litter or under bark, and come out only at night to hunt and feed on insects, snails, worms and other smaller centipedes. Most species inject venom into their prey and, although the bite may be painful to humans, it usually only results in localised swelling and irritation. Colloff et al. recently provided the first comprehensive key to Australian centipedes. The widespread genera Scolopendra, Cormocephalus and Ethmostigmus occur in the region, as does the introduced European house centipede (Scutigera coleoptrata).

Arachnids (Spiders, Scorpions and their Relatives)

Araneae (Spiders)

The region represents a meeting point for many South Australian arachnids. Here many species from the higher rainfall south-eastern Australian region overlap in distribution with those of the semi-arid north. Additionally there is an influx of species following the riverine habitat from the eastern states and probably the most obvious of these is the common large River Murray huntsman spider (Holconia murrayensis) (Sparassidae) associated with the river red gums that provide adequate shelter in the form of hollows, crevices or large loose slabs of bark. Other spiders found along the river as an extension from the east include the water spider (Dolomedes facetus) and the jumping spider Astia.

In general terms most spiders are divided into the Mygalomorphae in which the fangs lie parallel and the Araneomorphae in which the fangs of most species are pincer-like. The Araneomorphae are further divided into primitive and more advanced families. Raven et al. have provided an interactive key to the identification of spider families.

The Mygalomorphae are known for their long life span and the females are often not seen unless the burrows are excavated. Males are more likely to be encountered as they wander in search of a mate in times of high humidity, usually after rain. Of the Idiopidae, Blakistonia aurea occurs to the west of the river and at least one species of Aganippe is found in the southern area. Both species have a door to a deep burrow with that of Blakistonia being semi-circular with a crenulated lip to both door and burrow entrance, and shaped like a clam shell. Other trapdoor spiders known from the Murraylands include at least three species of mouse spiders (Actinopodidae) which are the only mygalomorphs where the male wanders during daylight hours. The red-headed mouse spider (Missulena insignis) is common, whereas M. rutraspina is found in the extreme north-eastern sector and the smaller M. dipsaca is found in the mallee belt through to Western Australia (Faulder 1995). These two species have a black or blue-black head and females of all species are stout, dark brown to black, while that of M. insignis can be much larger than the male. Another large black conspicuous species is the brush footed trapdoor (Idiommata scintillans), the male of which has a coat of silvery hairs on the head. The doors to their shallow burrows are well camouflaged and difficult to find.

Red-headed mouse spider (Missulena insignis) male, near Blanchetown SA.

Aname robusta (Nemesiidae) may have the honour of being the first spider described from the Riverland and Murraylands but its status is obscure. Although the type from Mannum is in the Australian Museum, no other specimens have been attributed to the species. A pale Aname species occurs in the mallee country west of Morgan and other similar species are known in the southern areas. Species of Kwonkan are known to occur throughout the region but burrows are hard to find as a collar of loose silk is pulled together during the day to cover the entrance. Other species of Nemesiidae may spin a thin sheet of silk over the burrow entrance but none have trapdoors. The black wishbone spider group is represented by one large species, Aname sp., and aggressive males may be encountered after rains inciting fear that they may be the infamous funnel-web spider. In the north- eastern sector the whistling or so called 'bird-eating' spider (Selenocosmia stirlingi) is locally common. Grasshoppers, other large insects, frogs, small lizards and the occasional mouse are likely to be a part of its diet as they are generally nocturnal.

Primitive araneomorph families found in the region include Segestriidae (Ariadna or Segestria) in which some species construct strong dense tubes of silk under the bark of trees with the base of the tube often firmly secured in a borer hole or crevice, while others live in silk-lined burrows in the ground. Of the Oonopidae, also known as six-eyed spiders, the pouched spider (Grymeus robertsi) is one of many species found in the area. All are of smooth appearance, yellow or red-brown in colour and are seldom seen due to their small size and habit of living under bark, rocks or in leaf litter.

River red gums provide an important habitat for many tree dwelling spiders found in the region and of these a number of species can be found in large numbers. Close inspection under slabs of old bark will reveal the small cryptic Wandella murrayensis (Filistatidae), a small spider with an exceptionally long life and the ability to go long periods with little food or moisture. Hollowed trunks as well as rocky cliffs provide an ideal habitat for the native daddy-long-legs spider (Wugigara kaurna) (Pholcidae), while the introduced species Holocnemus pluchei and Pholcus phalangioides, are more likely to be found in human habitation. Webs of the black house spider (Badumna insignis) (Desidae) can be seen in profusion on old red gums that have a thick layer of old bark at the base. Other members of Desidae found in the region include Forsterina spp. in sparse webbing under stones or logs, Phryganoporus davidleei which usually makes webs in the branches of shrubs, and P. candidus which is sometimes communal and produces dense webbing on the foliage of trees and can be a nuisance in orchards. The much smaller Callevopthalmus sp. makes a small web over a depression in the ground and the spider waits for prey in a sheltered crevice at one side. A number of very small desid species are found on tree trunks or in leaf litter, while a pale coloured species inhabits saline areas subject to ephemeral flooding.

Three species of the Nicodamidae, red and black spiders, spin sparse webs at the base of tree trunks or in similar sheltered positions. The largest is Nicodamus peregrinus and the long legged male is most strikingly coloured. Two smaller species are Ambicodamus leei and Durodamus yeni. Despite the red colour signifying a warning these spiders are considered harmless. The platform spider (Corasoides sp.) (Stiphidiidae), lives in a short burrow in the ground from which a tube of silk rises funnel-shaped into the undergrowth, where it is transformed into a large silken platform above which straggled silk threads act as a knockdown for unwary insects flying through.

The more advanced araneomorph spiders include two species of water spiders (Dolomedes) although one species, D. facetus, has only been recorded near the Victorian border. Wolf spiders of the genera Artoria, Trochosa and Venatrix commonly occur on the flood plains of the Murray River with V. speciosa and V. goyderi being found close to water. The common V. pseudospeciosa is often found in gardens that enjoy regular watering. Lycosa leuckartii is the largest common species to be encountered and often has makeshift burrows in the cracked soil of the northern part, but is replaced by L. gilberta in the southern part of the river floodplains. In drier areas away from the river wolf spiders are prevalent with at least 20 species known. Several species, one being Dingosa serrata, construct a palisade at the burrow entrance either of leaves, grass stems, small twigs or of silk and soil. Another smaller wolf spider, Venonia sp., makes horizontal sheet silk webs amongst leaf litter on the ground which are difficult to distinguish, except perhaps for their larger size, from those made by Callevopthalmus sp. (Desidae).

Other spiders which make webs on or near the ground include species of Steatoda (Theridiidae) and one introduced species, S. grossa, which is associated with human habitation. The tiny Hadrotarsus is also often found on the ground. The web of the redback (Latrodectus hasseltii) is almost a reverse of Corasoides with a funnelled retreat above and a tangle of sticky threads below to snare insects and even small reptiles moving on the ground. The grey house spider (Achaearanea sp.) is generally associated with buildings while several other species in the genus can be found in more natural situations, often on tree trunks. The related Euryopis sp. can be seen at night hanging on a silk line from the branches of tree trunks feeding on ants much larger than itself. On seizing an ant that is walking along the branch then dropping with it on a silk line, the ant is left powerless with nothing to grip onto.

Tamopsis fickerti (Hersiliidae) employs a different tactic and, on encountering prey, circles it rapidly playing out streams of silk from long spinnerets (hence the name 'two-tailed spiders') rendering the insect helpless. One member of the Thomisidae, Tmarus sp., appears to use the same technique as Euryopis while other family members, divided between crab spiders (Stephanopis and Tharpyna) and flower spiders (Sidymella and Diaea cruentata), are sit-and-wait predators. Stephanopis cambridgei (Thomisidae) is found under bark while a second species has been collected more often in leaf litter. Huntsmen spiders can similarly be divided with the sit-and-wait ambush method being employed by Isopeda leishmanni (south of Mannum), Isopedella cerussata (northern sector) and I. saundersi in mallee areas, while adjacent to the Victorian border is I.

frenchi. Delena cancerides is rarely seen as it hunts under large loose slabs of bark rather than on tree trunks. This habit is somewhat emulated by Holconia nigrigularis, a close relative of H. murrayensis, but which is restricted to the drier north-eastern corner of the region and more likely to be found in stands of large Casuarina trees. Pediana occidentalis can be found north of the river in mallee vegetation and an intricate pattern of drab colours ensues that it is well camouflaged against the bark. The more active badge huntsmen (Neosparassus spp.) may be seen at night moving over the foliage of low shrubs in search of prey. Two species common on the east coast, N. calligaster and N. punctatus, can be found in the southern Murraylands. Another six or seven species known in the area are undescribed. Several are recorded from only one or two localities, only one is found throughout the area and all are restricted to undisturbed natural vegetation. Keilira sparsomaculata, one of South Australia's smallest huntsmen, with a body length of less than 1 cm, is another active species occurring in heathland in the south.

Sac spiders (Clubionidae) are represented by two common genera. Pale sac spiders (Cheiracanthium) prefer foliage and seem to have a preference for prickly bushes or thistles, while Clubiona is found under the bark on trees and constructs short silken tube-like nests (Austin 1984). Also found on trees and shrubs are some members of Gnaphosoidea, a large group comprised of several families which include the Gnaphosidae, Galliellienidae (Meedo mullaroo and Oreo renmark, both of which are rarely collected) (Platnick 2002), Prodidomidae with unusual long mid ventral spinnerets, and Trochanteriidae, members of which are referred to as 'flat rock spiders' although the common species, Longrita millawa, can also be found under bark on tree trunks. Liocranidae is little known and one small species of Orthobula is recorded by an undescribed species. The notorious white-tailed spider (Lampona cylindrata) (Lamponidae), naturally occurs under bark or in sheltered rocky areas where it preys on other web building spiders such as the black house spider. There are a number of closely related but smaller relatives, Lampona gilles, L. braemar, Lamponina loftia, L. asperrima, Bigenditia millawa, Notosdipus dalby, Asadipus bucks and Prionosternum nitidiceps (Platnick 2000). The latter was described in 1909 from Western Australia and represents an element of the Riverland and Murraylands fauna that had a distribution once continuous across South Australia and into Western Australia. Aridification has now confined these once widespread species to the shrinking refuges that remain.

White-tailed spider (Lampona cylindrata).

The Gnaphosidae are well represented with numerous species of Encoptarthria and Hemicloea, and at least one species in both Anzacia and Eilica. Many other undescribed genera and species occur in the region. The extremely flattened species of Hemicloea are adapted to living between layers of bark but the other genera are found on the ground, under rocks or living in leaf litter.

Other spiders living on the ground include the large family of spotted ant spiders of which Neostorena victoria is a burrower and adorns its burrow entrance with a palisade of large leaves. The female is about twice the size of the male and is possibly the largest species in the family. Storena formosa is occasionally encountered under objects on the ground and, as with other members of the genus, has large conspicuous yellow spots on the abdomen. The much smaller Pentasteron intermedium and P. storosoides feed on small black ants, while Habronestes bradleyi is usually encountered close to nests of meat ants (*Iridomyrmex*). Little is known of the habits of other species including Cavasteron crassicalcar and C. atriceps, Holasteron aciculare, H. hirsti and H. spinosum, Australutica quarens, Zillimata sp. and many other undescribed species.

Numerous species of the closely related Zoridae and Miturgidae occur throughout the region but little is known about them. Zorids may be seen during the day as they move around cautiously through the leaf litter layer, but species of Miturga (*lined spiders*) are strictly nocturnal and make silken tube-like nests under stones, logs or even in dense low shrubs near the ground. The white and yellow spotted Supunna picta (*Corinnidae*) is often seen as they dart about in open areas even on sunny days. Other genera such as Poecilipta are more ant-like in appearance and may have metallic hues. An unnamed genus and species with a colourful reddish or even blue anterior body constructs short burrows in the ground with a palisade of silk-bound debris rising from the entrance.

Jumping spiders (*Salticidae*) may well be the most diverse and numerous group in the region, however most are small and grey or brown coloured and not readily seen. Ocrisiona melancholica, Helpis occidentalis, Servaea vestita, Breda jovialis and Clynotis severus are larger species commonly found under red gum bark. All may be seen hunting on tree trunks during the day. Breda jovialis has a distinctive yellow-gold patch on the abdomen and also preys on the black house spider. Small, flattened species of Holoplatys and Zebraplatys find their way into narrow crevices between the bark of mostly smaller eucalypts. One of South Australia's largest and brighter patterned jumping spiders, Sandalodes superbus, lives exclusively in rolled bark that has dislodged from a tree and been caught in shrubs below. Although recorded from both the northern and southern fringes of the Riverland and Murraylands, it no doubt would be present along the river valley. Lycidas is a large genus with a number of species in the area which are usually seen on the ground. The male of L. chrysomelas is brightly coloured and looks very unlike the drab female. Both sexes of Jotus, in which most species have a distinctive black stripe along the middle of the abdomen, are often seen during summer days in a variety of habitats close to the ground. A species of Grayenulla, though commonly collected in pitfall traps, is seldom seen and little is known of its habits. Another family of spiders that have jumping movements, particularly when disturbed, is Oxyopidae. Oxyopes molarius, O. amoenus and O. quadrifasciata are all of similar appearance and live in low herbaceous vegetation. The eggsac is constructed in the upper branches and the female sits astride it guarding against predators.

Many spiders are known to 'balloon' or travel across country on wind-carried silken threads. This practice is not restricted to young spiderlings dispersing from the maternal nest but includes adult midget spiders (*Laetesia and Dunedinia spp.*) (*Linyphiidae*) and even small wolf spiders of the genus Artoria. For those species the Murray River is certainly no barrier to dispersal to new suitable habitats. It is likely that the orb-weavers (*Araneidae*) are also adept at ballooning as at least the larger common garden orb-weavers (*Eriophora heroine*) appears to have established in New Zealand. Along with E. biapicata, this species is found in natural areas or cultivated paddocks, although E. biapicata prefers to make a web higher up off the ground whereas E. heroine will often have a web lower to the ground in grass. Grass or low shrubs is also preferred by the tear-drop spider (*Argiope protensa and Gea theridioides*). while the bird-dropping spider (*Celaenia kinbergi*) is more likely to be found hanging in leafy foliage of fruit trees. Orb-weaver species common along the riverbanks that rely on a more humid atmosphere are Araneus eburnis, the spiny spider (*Austracantha minax*) which can be seen in semi-communal webs in low shrubs, and the well camouflaged Larinia phthisica which makes a small orb web amongst the long grass and sedges. Araneus usualis is normally found on dead trees or branches particularly near water. At least two species of the wrap-around spider (*Dolophones*) are found in the branches of trees and rest during the day with their legs partly tucked beneath the peculiar-shaped body resembling a node on a small branch. Carepalxis and a species of Poltys have a similar habit but the abdomen of the latter is much higher and its unusual shape can vary considerably. Numerous orb-weavers of the family Tetragnathidae abound. The long-jawed spiders most often build their orb-webs over water amongst rushes. The golden-orb weaver (*Nephila edulis*) is widespread and can be very common in some seasons, while the equally common leaf curlers (Phonognatha graeffei and P. melania) are less obvious. In borer holes of dead trees P. dimidiata shelters during the day emerging at night to make a web. The tiny Pararchaea (*Pararchaeidae*) has been collected from Scott Creek near Morgan. When hunting, the fang bases (*chelicerae*) are held at right angles to the body and on encountering prey, the fangs likely close upon it rapidly and with some force.

Opiliones

The paucity of Opiliones in museum collections from the Riverland and Murraylands suggests they may be rare. In fact there have been just four specimens collected from the area. However, it may reflect more their cryptic nature as those four specimens represent three species and three families. The short legged harvestman Dampetrus gracilis (*Triaenonychidae*) was previously known only from Red Cliffs, Victoria, and following a recent survey near Chowilla is now a new record for South Australia. Two long-legged Opiliones are known, Megalopsalis sp. (*Megalopsalididae*) and the tiny fragile looking Ballarra sp. (*Neopilionidae*) which is considered to be restricted to Gondwanan continental fragments.

Acari (Mites and Ticks)

Although many mites are predatory, there are also many species of plant feeders, fungivores, saprophytes, pollen and nectar feeders, microbial filter feeders, and internal and external parasites on a wide range of vertebrate and invertebrate hosts.

Most mites commonly encountered in extensive and intensive agricultural systems are introduced and are usually considered pest species. Among those commonly found in horticultural crops are the pest species two-spotted mite (*Tetranychus urticae*) and bryobia mite (*Bryobia rubriocolus*). Two-spotted mite has a large host range including apple, pear, peach, plum, prune, strawberry, bean, cucurbits, as well as a very wide host range among woody and herbaceous crops and weeds. Both nymphs and adults feed by piercing and sucking the lower leaf surfaces, leading to the development of yellow-green mottling and loss of vigour and yield. The beneficial mite (*Phtyoseiulus persimilis*) is common in horticultural regions where it has been introduced. It is a voracious feeder on all stages of two-spotted mite. Bryobia mite can be found on almonds, prunes, plums, pears, apples and apricots, although it is most damaging to almonds, prunes and sometimes pears. This mite damages epidermal cells by piercing the cells and sucking out the contents. The upper leaf surfaces have a stippled green grey appearance. The bronzing typical of two-spotted mite is not produced.

Eriophyid mites (*Atopomelidae*), grapeleaf blister mite (*Colomerus vitis*) grapeleaf rust mite (*Calepitrimerus vitis*), and bunch mite (*Brevipalpus californicus*) (*Tenuipalpidae*) can all be found on grape vines in the region. Grapeleaf blister mite feeds on the undersides of leaves producing yellow areas which develop into raised blister-like swellings on the upper surface. The cavities on the underside of the blister are filled with felt-like hairs. The life cycle of the grapeleaf rust mite is similar except that it mainly occurs on the upper side of the leaves, whereas bunch mites are most commonly found in the bunches and on the undersides of adjacent leaves.

A very common pasture and crop pest in the region is redlegged earth mite (*Halotodeus destructor*) (*Penthaleidae*). This mite feeds on a wide range of field crops, vegetables, flowers, pastures and weeds. It is mainly a pest of broad-leafed plants and feeds by puncturing plant cells and sucking up the sap, producing a mottled effect which gives the leaf a grey or white appearance.

Not all mites in the region are pest species. For example, several families including Erythraeidae and Trombidiidae have larvae that are parasitic on insects, and have terrestrial adults that are often called red velvet mites, including (*Erythrites reginae*) which is commonly seen in crops and pastures in the region.

The order Sarcoptiformes includes soil mites in the Suborder Oribatida, commonly known as oribatid mites. These are mostly slow-moving, heavily sclerotised mites that are most abundant and diverse in the upper layers of soil and associated organic litter. They feed on dead plant material and associated microfungi. The Sarcoptiformes also includes the Suborder Astigmata, including the common flour mites and cheese mites that infest stored food and which damage stored grain, the house dust mites that cause allergy and asthma in humans, and a wide range of fungivorous and saprophytic species in soil. Many of the Astigmata have a specialised immature form known as the hypopus, which attaches itself to insects for dispersal. The Astigmata also includes feather mites of birds. Domrow catalogued the species of Astigmata that parasitise Australian vertebrates or occur on their bodies. These include the mites that cause such diseases as scabies (*Sarcoptes scabiei*), scaly leg of chickens (*Knemidokoptes mutans*) and several types of mange in domestic animals (e. g. the ear mite of dogs and cats, Otodectes cynotis).

There are about 80 known species of hard ticks, Order Ixodida, in Australia. They are all blood-feeding parasites of mammals, reptiles and birds, and are characterised by mouthparts with a large number of backward-directed hooks that are inserted into the host for feeding. Of these, the reptile tick (*Aponomma hydrosauri*), can be found parasitising sleepy lizards (*Trachydosaurus rugosus*) and bluetongues (*Tiliqua spp.*) on the north bank of the Murray River but has not yet been found more than a few kilometres beyond it. The reptile tick, Amblyomma limbatum, also occurs in the region north of the 400-mm isohyet. Other ticks from reptiles include the goanna ticks Ap. jimbriatum and Ap. undatum, and the common marsupial tick (*Ixodes tasmani*).

Scorpions

The scorpion fauna of southern Australia comprises representatives of three families, Buthidae, Bothriuridae and Urodacidae, all of which are found in the Riverland and Murraylands. Scorpions are active predators, emerging from their burrows at night to prey upon other invertebrates. They give birth to live young which are carried on their mothers' back for several weeks prior to dispersal.

The buthid fauna of the Murraylands comprises several species of Lychas and a single species of Isometroides, which are characterized by their slender pedipalps and generally mottled yellow-brown colour. Species of Isometroides have a highly specialised diet, as they preferentially hunt burrow-dwelling spiders such as trap-door and wolf spiders. The Bothriuridae contains the single genus Cercophonius which tend to be best represented in forested ecosystems. Cercophonius kershawi occurs in the Murraylands in leaf litter of mallee trees. The Urodacidae with the sole genus Urodacus are restricted to mainland Australia, and include the largest scorpions in the country. Urodacus manicatus lives in shallow depressions under rocks in south-eastern Australia, including the Murraylands. The desert-dwelling U. yaschenkoi has been recorded near Renmark, Berri and adjacent regions of Victoria and New South Wales, where it digs deep burrows in the soft soil of dunes. Urodacus armatus digs shallower burrows on the dune base and inter- dune swales.

Scorpion – *Urodacus yaschenkoi.*

Pseudoscorpions

These are small arachnids, less than 1 cm in length, which resemble scorpions but lack the elongate tail and terminal sting. Although fairly common in most terrestrial ecosystems, they are rarely observed due to their secretive habits. Most research on these animals in South Australia has concentrated on cavernicolous species from the South-East and the Nullarbor Plain. The regional fauna is represented by just a few families. Several species are commonly found under the bark of trees, particularly eucalypts, including the common Oratemnus punctatus (Atemnidae), several species of Conicochernes (Chernetidae) and at least two species of Synsphyronus, S. dewae and S. niger (Garypidae). Others occur closer to the ground where they occupy leaf litter and soil habitats. The best known examples include the widespread Geogarypus taylori (Geogarypidae), which occurs over much of southern Australia and Synsphyronus mimulus, but other families such as the Cheliferidae, Olpiidae and Chthoniidae also occur in the area.

Insects

Introduced and Native Insect Pests

Introduced insects are widespread and common in the region, with most being able to be treated conveniently as occurring in either intensive or extensive agricultural systems. Whilst some of the introduced insects such as the cabbage white butterfly (*Pieris rapae*) and honey bee (*Apis mellifera*) are ubiquitous, most are confined to the agricultural areas, tending not to be so common in natural environments where plant and animal hosts often do not occur.

The intensive agricultural systems include crops such as citrus, stone fruits, grape vines, and vegetables, and also irrigated pastures and some dairying, particularly along the lower reaches of the Murray River. In comparison, extensive systems are mostly dryland where cereals, as well as crops such as canola and field peas, and domestic stock such as sheep and cattle, are farmed.

Introduced Insects Associated with Intensive Systems

The majority of introduced insects found in intensive systems are considered pest species. Many of these can also be found in extensive systems.

There are a large number of introduced pest Hemiptera, and among the more significant are: mealybugs such as citrus mealybug (*Planococcus citri*) and the longtailed mealy bug (*Pseudococcus longispinus*); scale insects such as soft brown scale (*Coccus hesperidum*), grapevine scale (Parthenolecanium persicae), black scale (*Saissetia oleae*), red scale (Aonidiella aurantii) and yellow scale (*A. citrina*); aphids such as the black peach aphid (*Brachycaudus persicae*), green peach aphid (*Myzus persicae*), black citrus aphid (*Toxoptera aurantii and T. citricida*), and cabbage aphid (*Brevicoryne brassicae*); and true bugs such as the green vegetable bug (*Nezara viridula*).

Several thrip species (*Thysanoptera*) are significant in the region, particularly the western flower thrips (*Frankliniella occidentalis*). This thrip can attack more than 250 plant species including many vegetable and flower crops, and many weed species. Preferred hosts include capsicums, lettuce and chrysanthemums. It causes flecking, silvering or deformation of flowers, growing tips, young foliage, stems and fruit, and can transmit tomato spotted wilt virus. Other species include Kelly's citrus thrip (Pezothrips kellyanus) and tomato thrips (*Frankliniella schultzei*).

Fuller's rose weevil (*Asynonychus cervinus*), the vegetable weevil (*Listroderes difficilis*) and white-fringed weevil (*Naupactus leucoloma*) are among the more important introduced pest beetles. The economic damage caused by Fuller's rose weevil is mainly confined to citrus, but a wide range of plants such as olive, ornamentals such as roses, some vegetables, as well as many weeds are attacked. The vegetable weevil attacks cruciferous plants, especially cabbages and turnips, celery, carrots, lettuce, silver beet and spinach, winter growing flowers such as pansies, and weeds such as marshmallow and capeweed are favoured by the larvae. Adults also damage crucifers, potatoes and tomatoes. Adult whitefringed weevils cannot fly and emerge from the soil in summer. The larvae of this weevil are highly polyphagous and chew the roots of a wide range of plants including clovers, lucerne, peas, water melons, tomatoes, potatoes, rye grass, wheat, brassicas and onions. The African black beetle (*or lawn beetle*) (*Heteronychus arator*) (*Scarabaeidae*) is also widespread and common. It has a broad host range including maize, lawns, tomato, cabbage, potato, rhubarb and various garden flowers, and other market garden plants, and a wide range of grasses, but not legumes. Most damage is caused by the last larval stage feeding on roots, while young larvae tend to feed on soil organic matter. Adult beetles chew on stems just below ground level.

A number of introduced pest moths and butterflies can be found in the region, among them cabbage white butterfly (*P. rapae*), oriental fruit moth (*Grapholita molesta*), codling moth (*Cydia pomonella*) and diamondback or cabbage moth (*Plutella xylostella*). The cabbage white butterfly is perhaps the most obvious, and its host plants include cruciferous plants, especially cabbages, cauliflowers, brussels sprouts, broccoli, turnips, radish, and mustard, as well as various weeds and garden flowers.

A wide range of beneficial insects has been introduced into the region as biological control agents. Many of these are minute parasitic wasps such as Aphytis melinus (*Aphelinidae*), which is a parasite of armoured scale insects, particularly red scale in citrus. It lays eggs under the scale cover, and the developing larva feeds on the scale insects killing it, and eventually emerging as another adult wasp.

Whitefringed weevil adult (Naupactus leucoloma).

Introduced Insects Associated with Extensive Systems

The lucerne flea (*Sminthurus viridis*) (*Collembola: Sminthuridae*) is a native of Mediterranean areas and is well established in southern Australia. Hosts include medics, clovers, beans, cereals and a wide range of mainly broad-leaved crops and weeds. Lucerne flea causes damage by chewing, and results in the leaf surface having an exposed 'membrane' or epidermis giving a characteristic 'window' on the remote surface. Minute black specks of excreta are also present.

The African mole cricket (*Gryllotalpa africana*) (*Gryllotalpidae*) are quite distinctive. They are velvety brown, thick-set, about 25-40 mm long, and have strong forelegs well adapted for burrowing. The fore wings are very short, but they can fly quite strongly. The hind wings are large and fan-like when unfolded. Mole crickets damage grasses and sometimes tomatoes, vegetable and flower crops by tunnelling through the roots. Succulent stems may be chewed at ground level. They may occasionally chew into root crops such as carrots and turnips.

Among the various aphid (*Aphididae*) pests of extensive crops are bluegreen aphid (*Acyrthosiphon kondoi*), which is a native of Korea and Japan. It is chiefly a pest of lucerne, subterranean clover and medics but also peas, lentils, yellow lupins and melilotus can be hosts. The aphids suck sap from the terminal areas of plants; those parts that are tender and succulent. Stems often become severely stunted with small leaves and shortened internodes. The leaves turn yellow, curl and may drop off in severe infestations. Cowpea aphid (*Aphis craccivora*), oat or wheat aphid (Rhopalosiphum padi), pea aphid (Acyrthosiphon pisum) and spotted alfalfa aphid (Therioaphis trifolii f. maculata) are also common in extensive crops.

Various introduced beetles are pests. These include sitona weevil (Sitona discoideus) (Curculionidae), which was first recorded as a pest in New South Wales in the late 1950s, but now occurs across southern Australia, including Tasmania. It is a significant pest of annual medics and lucerne. The pea weevil (Bruchus pisorum) which, despite its common name, is not a weevil, but rather a member of the Bruchidae. It is a cosmopolitan pest of field peas, and was first recorded in South Australia in 1964.

A range of storage insect pests are commonly found in and around grain storage areas on farms, and occasionally in people's houses. These include the lesser grain borer (Rhyzopertha dominica), granary weevil (Sitophilus granarius), rice weevil (S. oryzae), and several species of spider beetles (Anobiidae) including the shiny spider beetle (Mezium affine). The Indian meal moth (Plodia interpunctella) is also commonplace.

Introduced Insects Associated with Domestick Stock

As well as pests associated with crops and pastures, there is a wide range of introduced pest species associated with domestic stock. These include fleas such as the cat flea (Ctenocephalides felis), dog flea (C. canis), and poultry stickfast flea (Echidnophaga gallinacea). A number of important lice are also found in the region as ectoparasites of domestic stock and include shortnosed cattle louse (Haematopinus eurysternus), longnosed cattle louse (Linognathus vituli), and sheep body louse (Bovicola ovis).

The sheep nasal botfly (Oestrus ovis) (Oestridae), is common and widespread. The female fly lays batches of about 50 newly hatched larvae in or on the nostrils of the sheep; the larvae crawl onto the mucous membrane of the nasal passage, where they remain for at least two weeks. Development of the first instar larva may be delayed for one to nine months, and plays a role in the over-wintering cycle. The pupal stage occurs in the ground and lasts one to two months depending upon temperature.

Sneezing and a mucopurulent nasal discharge (snotty nose) may be a result of botfly larvae. The main effects for the host are persistent annoyance and associated debility.

Introduced beneficial insects are found in the region include dung beetles, such as Aphodius lividus, which have been introduced to supplement the native dung beetles such as the threehorned dung beetle or Kershaw's burying beetle (Onthophagus mniszechi). As well, various parasitic wasps such as lucerne aphid parasite (Aphidius ervi), are widespread and common.

Species Commonly Associated with Humans and their Domestic Pets

Apart from those insects included above which can often be found in home gardens, there are a range of insects commonly associated with humans, either in their homes and gardens, or actually attacking humans or their pets.

Several cockroach species (Blattodea) are common in the region, particularly the German cockroach (Blatta germanica), Oriental cockroach (B. orientalis), American cockroach (Periplaneta americana), and the native Australian cockroach (P. australasiae). The latter species has been spread by humans so that it is found extensively throughout Australia as well as overseas.

Termites (Isoptera) play an important role in nutrient recycling, but the feeding of some species may be of considerable economic importance with damage occurring to timber in buildings and fences. In the region, common pest termites include Coptotermes acinaciformis (Rhinotermitidae), with nests which are subterranean or in stumps or hollowed trunks of dead or living trees, and Nasutitermes exitiosus, with largely subterranean nests. Other termites include Cryptotermes spp., Coptotermes frenchi and Heterotermes ferox.

The most widespread introduced earwig (Dermaptera) is the European earwig (Forficula auriculari). Earwigs are rather cryptic, small to medium sized insects distinguished from other insects by a pair of forcep- or pincer-like cerci at the end of the abdomen. Forficula auricularia can be a pest of gardens and vegetable crops, and is relatively common around homes. In other areas of the state, it has caused damage to the seedlings of emerging canola crops.

The rose aphid (Macrosiphum rosae) is the most common insect pest of roses in Australia. Aphids damage plants by sucking sap from plant tissues using highly specialised sucking mouthparts. The usual symptoms of damage are distortion of new leaves and flowers. Aphids also excrete large amounts of honeydew, which can cover the plant, resulting in the growth of sooty moulds and the consequent reduction in photosynthesis and aesthetic appeal. Severe infestations can result in defoliation of the plant and loss of the flowers. As with most aphids, they can breed very rapidly and build up vast numbers, especially in warm, humid weather. The rose aphid parasitoid (Aphidius rosae), has been successfully introduced from Italy as a biological control agent.

Recently, bed bugs (Cimex lectularius) (Cimicidae), have undergone a dramatic resurgence and worldwide there are reports of increasing numbers of infestations. Bed bugs are wingless blood- sucking insects, oval in shape and 4-5 mm long when fully grown. They are rust brown in colour and change to a deeper red-brown following a blood meal on their human host. Bed bugs are dorsoventrally flattened and being thin means that they can hide in narrow cracks and crevices, making detection often very difficult. The mouthparts of bed bugs are especially adapted for piercing skin and sucking blood, and each nymphal stage requires a blood meal. Like most blood sucking arthropods, they inject saliva during feeding, which has anticoagulant properties. Bed bugs respond to warmth and carbon dioxide of a host and quickly locate a suitable feeding site. Most blood feeding occurs at night, and they generally seek shelter during the day and become inactive while digesting the blood meal. They can survive for long periods without feeding. While their preferred host is human, they will feed on wide variety of other warm-blooded animals including rodents, rabbits, bats, and even birds.

Many species of Muscidae are of economic importance including the cosmopolitan housefly (Musca domestica) which is found in association with humans or activities of humans, and was probably introduced with European settlement. It is commonly found around homes and also pig and poultry enterprises, horse stables, and farms. Not only are they a nuisance, but they also can transport disease-causing organisms. Excessive fly populations are obnoxious, and they can pose a public health problem. Other important species include the bush fly (M. vetustissima), and the cosmopolitan stable fly (Stomoxys calcitrans).

The most common blowfly (Calliphoridae) of the region is the Australian sheep blowfly (Lucilia cuprina), which is responsible for initiating most blowfly strikes in sheep. It is, however, not endemic to Australia and was accidentally introduced into Western Australia in the late 1860s, most likely from South Africa.

Australia's early European settlers introduced honey bees (A. mellifera) to ensure a good supply of honey. They are now wild throughout Australia, including the Riverland and Murraylands. Honey bees play an important role as pollinators of crops and wild flowers, but some wild flowers have suffered from the presence of honey bees as these flowers can only be effectively pollinated by native bees or birds. As well, by building their hives in places such as hollow tree limbs, birds and small mammals lose places in which to shelter and construct their own nests. Honey bee stings cause intense local pain and swelling, and a severe allergic reaction may occur in some people.

The European wasp (Vespula germanica) is native to Europe, North Africa and temperate Asia, and reached the Australian mainland (Melbourne) in 1977. It has since spread rapidly, and can now be found in many towns and cities in south-eastern Australia. Nests are located where shelter is available, with common locations being retaining walls, tree hollows, wall cavities, or underground. The nest is made of grey papier mache-like material. If a European wasp is aggravated it may sting, but unlike the honey bee, it can sting multiple times. Some people are allergic to the stings. Through sheer numbers, the predation of European wasps on native insects and honey bees can be deleterious but they can also affect native animals (including birds) through competition for food. The European wasp is sometimes confused with native paper wasps, Polistes spp., which are very similar in colour but have antennae that contain yellow, orange or brown markings as well as black. European wasps have antennae that are entirely black. Their nests are also distinctive, being a single 'honeycomb-like' layer which is usually horizontal. If 'honeycomb' cells can be seen it is unlikely to be a European wasp nest.

Native Insects that have Pest Status

When environmental conditions are right, native insects can build up in sufficient numbers to become pest species. In intensive systems, this may be a continual problem (i.e. the insect is considered a key pest) or if it occurs spasmodically, then the pest is considered an occasional pest.

The native plague thrips (Thrips imaginis) (Thripidae), has been recorded in all states in Australia, and has a host range that includes pears, grapes, peaches, strawberries, roses, and some vegetables. When numerous, they may be found in just about any flowers. Both nymphs and adults cluster inside the flowers and buds where they feed by rasping the surfaces and sucking the exuding sap.

The Australian plague locust (Chortoicetes terminifera) (Acrididae) is distributed widely across mainland Australia. Certain conditions lead to migration southward into areas not normally occupied, for example, cereal areas of South Australia. The locust has a wide host range of native grasses, but also attacks cereals, pastures and a range of horticultural crops and weeds. Other native acridid grasshoppers and locusts found in the region include the small plague grasshopper (Austroicetes crucita) and wingless grasshopper (Phaulacridium vittatum).

The black field cricket (Teleogryllus commodus) (Orthoptera: Gryllidae), is found in all southern Australian states, especially where soils are heavy and tend to crack on drying. Young plants of many types may be occasionally damaged. Plants attacked include ryegrass, lucerne, clovers, oat, maize, potato, bean, carrot, strawberry, cotton and sunflower.

Cottonycushion scale (Icerya purchasi) (Margarodidae), is considered to be native to Australia, although it was originally described in 1879 from specimens found in New Zealand. Its host range includes citrus, Acacia spp., climbing fig (Ficus spp.), roses, laburnum, and mulberry. In South Australia, control measures are seldom necessary, as this insect is kept in check by the predacious vedalia ladybird (Rodolia cardinalis) and their larvae, and is also parasitised by small flies (Cryptochaetum iceryae).

A number of Miridae occur in the region. Although their feeding behavior is variable, some species are phytophagous and may be pests of crops such as apples, lucerne and stone fruits (e.g. green mirid, Creontiades dilutus), but many are beneficial as predators of small soft-bodied insects or insect eggs.

Rutherglen bug (Nysius vinitor) (Lygaediae) is found in all Australian states. It has a broad host range including sunflowers, potatoes, stone and pome fruits, young citrus, grapevines, strawberries, crucifers, grain sorghum, linseed, and a range of weeds as well as native plants. Both nymphs and adults feed by piercing and sucking sap from young tissues that may cause wilting, and blackening of young buds.

Another native insect that has made the transition from native plant hosts to crops is the crusader bug (Mictis profana). This insect is found in all Australian states. Native food plants include acacias and sennas. It also attacks citrus, grapes, roses and wisteria. Cultivated sennas, cassias, and wattles are frequently attacked. The closely related Amorbus bug causes similar wilting of the terminal growing tips, but this species, although similar in size to the crusader bug, lacks the distinct pale 'cross' on the hemelytra (fore wings) and is restricted to Eucalyptus.

Among the commonly found Pyrrhocoridae of the region is the harlequin bug (Dindymus versicolor) which may attack a wide range of crop and ornamental plants.

The spined citrus bug (Biprorulus bibax) is an invader from the eastern states. Both adults and nymphs cause damage to citrus by piercing and sucking with their mouthparts. The result of this damage is a brown stain and gumming on the fruit rind, and premature fruit drop in young fruit or internal damage in mature fruit.

Of the blowflies (Calliphoridae), the hairymaggot blowfly (Chrysomya rufifacies) is common, and can cause secondary strike of sheep after initiation by L. cuprina, although it generally breeds in animal carcases. Other blowflies include the lesser brown blowfly (Calliphora augur) and the eastern goldenhaired blowfly (C. stygia).

Among the beetles, the spinetailed weevil (Desiantha caudata) (Curculionidae), can be found attacking grasses, including cereals and subterranean clover, and has been reported on other plants including peas and fruit trees.

False wireworms and wireworms from the beetle families Elateridae and Tenebrionidae may be found attacking a wide range of grasses including cereals, as well as other plants including oil seed crops, grain legumes and vegetables.

Various cockchafers (Scarabaeidae) including the blackheaded pasture cockchafer (Aphodius tasmaniae), redheaded pasture cockchafer (Adoryphorus couloni), and the honey beetle (Heteronyx flavus), are all commonplace in the region. The blackheaded pasture cockchafer can be found in pastures and lawns, including bowling greens and golf courses. Dung of horses, cattle and sheep is the preferred food of adults and larvae. It has also been recorded as attacking wheat, where most damage is caused by the third instar larvae. Their feeding can seriously reduce the quantity of available feed during winter and spring. The larvae forage at night culling off pasture plants near ground level and placing pieces within their tunnels. Foraging occurs whilst the soil surface is moist following rain showers. With the redheaded pasture cockchafer, feeding on the roots of pasture plants results in roots being chopped off just below the soil surface. Grazing stock feeding on these plants simply uproot them. Various birds (e.g. ibis) feeding on the larvae will also uproot the pasture.

Several moths of the family Noctuidae are significant pests in the region. The native budworm (Helicoverpa punctigera) can destroy flowers, buds and fruit of tomatoes, maize, lupins, sorghum, beans, crucifers, sunflowers, clovers, medics, and many other field crops, pastures and vegetables, as well as many weeds and garden flowers. The grapevine moth (Phalaenoides glycine) can be found attacking grapes, glory vine, fuchsias, Virginia creeper and some native plants. The build-up of the grapevine moth has been satisfactorily checked by predators on many properties in South Australia. The most common of these is the spined predatory shield bug (Oechalia schellenbergii) which spears the caterpillar with its proboscis and sucks out the body fluids. Other noctuids that are important include various armyworms, cutworms and loopers.

The lightbrown apple moth (Epiphyas postvittana) (Tortricidae) can be found attacking almost all cultivated fruit crops, herbs, ornamentals, Acacia and most legumes, as well as a wide range of weeds, including capeweed and docks. The caterpillars are usually pale green with a brown head and grow to 12-20 mm long, moving actively over the host plant and sometimes hanging by silken threads. They tunnel into the undersides of the leaves under a cover of fine webbing or web together leaves and buds whilst eating them away. They may also attack fruits that are in contact with each other or with leaves. When disturbed the caterpillars wriggle actively backwards or forwards out of their shelters. Caterpillars skeletonise the foliage, removing irregular areas to leave a ragged appearance. Leaf margins are often rolled over. Damage to fruit is usually superficial but can provide an entry point for rotting organisms. Damage usually occurs where two fruits touch, where a leaf has been pulled down over a fruit or in the cavity around the stalk. The larvae always feed under shelter.

(left) native budworm (Helicoverpa punctigera) larvae feeding on lupins;
9 (right) lightbrown apple moth (Epiphyas postvittana).

The grapevine hawk moth (Hippotion celerio) and the vine hawk moth (Theretra oldenlandiae) (Sphingidae) also occur widely in the region, and are occasional pests of grapes. Other sphingid moths are occasionally reported from other plant species. The caterpillars consume the whole leaf and, if present in high numbers, may quickly defoliate the vines.

The larvae of the small citrus butterfly or dainty swallowtail (Papilio anactus) (Papilionidae) can be found attacking new foliage growth of citrus and many other plants in the family Rutaceae. The large citrus butterfly (Papilio aegus aegus) which is generally confined to New South Wales and Queensland, has been intermittently observed in the region. For example, in 1973-74, when significantly higher rainfall and more humid conditions were recorded, many butterflies were reported from the Upper Murray, Adelaide Plains and Mount Lofty Ranges, but the butterfly has not become established in South Australia.

Native Insects in Natural Habitats

Apart from the aquatic/semi-aquatic insects described in Chapter 10, the Riverland and Murraylands have an enormous diversity of native insects which are mostly restricted to natural habitats. Given the very large numbers of taxa involved, we have not attempted to cover all orders, but have provided examples to the most common groups.

Minor Orders

Collembola, Diplura and Protura, although not insects, are all hexapods, that is, they have three pairs of legs. Of these, Collembola (springtails) are the most commonly encountered in soil and leaf litter. Most species of springtails feed on decaying vegetable matter, although fungi, algae and lichens are also an important food source for many species. Some species feed on decaying animal matter, such as earthworms, dead flies or other springtails. One common springtail is the purplish coloured Hypogastrura spp. which is often seen in large numbers on the surface of pools after rain.

Apart from the introduced cockroaches detailed above, native species include the brightly coloured Ellipsidion australe (Blattidae), often seen on flowers, and Laxta granicollis and Calolampra irrorata (Blaberidae), both of which can be found under bark and in leaf litter. In the latter two species, it is only the males that are winged.

Among the commonly encountered mantids (Mantodea) are Orthodera ministralis (green or garden mantid), Paraoxypilus tasmaniensis (Amorphoscelidae) and Archimantis spp. (large brown mantids). All three have broad distributions outside of the Riverland and Murraylands.

Several earwig species (Dermaptera) are found in the region. The common brown earwig (Labidura truncata) occurs, especially in sandy habitats.

Apart from the locusts and grasshoppers (Orthoptera) detailed above, many other katydids, crickets, grasshoppers and locusts occur in the region. Conspicuous among these are the inland katydid (Caedicia simplex) and the yellow winged locust (Gastrimargus musicus). In a recent survey of Calperum Station, three new species of raspy crickets (Gryllacrididae) were collected.

Thysanoptera (thrips) are small insects which can be leaf-, flower-, pollen-, and fungal- feeding, although some are carnivorous on small arthropods. Apart from the pest thrips mentioned above, there are many other species from the region including the giant thrips (Idolothrips spectrum) and Haplothrips victoriensis, a predatory thrips.

Although several families of Neuroptera occur in the region, the most commonly seen are Hemerobiidae (brown lacewings), Chrysopidae (green lacewings) and Myrmelontidae (antlions). The brown lacewing (Micromus tasmaniae), is very common; it feeds on aphids on cereal, pastures and vegetable crops. Green lacewings lay their eggs on stalks for protection, and all life-stages are common. Their larvae are generalist predators, especially of hemipterans such as aphids and scales, and many species cover themselves with debris (including the exoskeletons of their prey). Antlions are mostly ground dwelling, and the larvae of some species construct conical pits as traps for their prey. These pits can often be found in areas protected from rain, such as under raised logs, under cliff ledges along the Murray River, or under or around buildings. Green lacewing and antlion adults are often attracted to lights.

Hemiptera (Bugs, Leafhoppers, Cicadas, Aphids, Scale Insects and Lerps)

The most characteristic feature of the Hemiptera is the structure of the mouthparts which are adapted to extract the liquid contents of plants or animal prey by suction. They are mostly terrestrial and phytophagous (plant feeding), although the suborder Heteroptera includes groups which are carnivorous (e.g. assassin bugs, Reduviidae), and others which show varying degrees of adaptation to aquatic environments.

(left) Ant lion pits, Blanchetown SA; 11 (right) Lerp of Glycaspis brimblecombei.

Psylloidea (jumping plant lice, lerp insects) are widespread and common in the Riverland and Murraylands. Among these is the white lace lerp (Cardiaspina albitextura), which develops regular outbreaks on the river red gum, E. camaldulensis. Similar outbreaks by the closely related C. densitexta occur on pink gum, E. fasciculosa, in the South-East, upper South-East and Murraylands. Other species found on river red gums include Glycaspis brimblecombei, which produces edible, sugary conical lerps, Creiis corniculatus which forms inedible horn-shaped lerps and C. costatus which forms distinctly flattened, white shell-shaped lerps. There are many species that do not produce lerps, most commonly Acizzia spp. on Acacia and Cassia, and Trioza spp. on false sandalwood (Myoporum platycarpum) and Casuarina pauper and various species of Allocasuarina. The free-living mistletoe psyllids Acizzia loranthaceae and A. amyemae feed on the mistletoe Amyema pendulum which commonly infests E. camaldulensis and black box (E. largiflorens) in the region.

Common eriococcids (Eriococcidae) found in the region include the gumtree scale (Eriococcus coriaceus) which occurs on a range of eucalypts, and Cylindrococcus spiniferus which causes galls on Allocasuarina spp. Several species of Apiomorpha, in which the females form spectacular species-specific galls on eucalypts, occur. Apiomorpha karschi is often found on E. camaldulensis.

The woody female gall occurs on stems, with mature galls often aggregating to form large woody swellings which can easily exceed 10 cm in diameter, whereas the small tubular male galls (up to 4.2 mm long) are mostly found massed together on the upper surface of leaves. Other common species include A. calycina, in which the female gall is somewhat cylindrical with a truncate apex. It occurs on E. dumosa (white mallee), E. gracilis (yorrell), and E. oleosa (red mallee) in the region. The female of A. malleeacola forms an ovoid gall which can be found on E. dumosa, E. oleosa, and E. socialis (red mallee). The female gall of A. munita is a distinctive four-sided gall usually with long extensions at each corner of the gall, and occurs on a wide range of eucalypt species. The female gall of A. strombylosa has a rough surface covered with small protuberances and resembles the seed cones of Allocasuarina/Casuarina. As the species name suggests, the female galls of A. urnalis are urn-shaped, and in the region occur on E. dumosa, E. gracilis and E. largiflorens (black box).

Whilst many scale insects occur, the armoured scales (Diaspididae) are perhaps more commonly observed. They include Maskellia globosa which is unusual in that it is a gall-former and does not produce an external scale. The male galls are small and horn-like and occur either on leaves or associated with the larger rounded female galls on twigs of various eucalypts.

Although cicadas (Cicadidae) are predominantly tropical/subtropical, a number of species have distributions which extend into the Riverland and Murraylands. These include the eastern sand-grinder (Arenopsaltria nubivena), the black squeaker (Pauropsalta encastica), and the cherrynose (Macrotristria angularis). Adults of the latter species have been collected on Callitris and Allocasuarina/Casuarina. An interesting species is Marteena rubricanta which is known from only three disjunct areas; Ravensthorpe WA, Morgan SA and Eurabalong western NSW. The adults, and possibly the nymphs, appear to be associated with mallee species. Several Cicadetta spp., including C. adelaida, can also be found along the Murray River.

A number of Cicadellidae (leafhoppers) are particularly characteristic of the mallee areas of South Australia (as well as WA and western NSW). The Subfamily Austroagalloidinae is endemic to Australia, and comprises eight described species, all on Eucalyptus, with two endemic to the Karoonda district. Deltocephalinae: Goniagnathini are arid-zone adapted species, with two of the five Australian species described from Keith. Several ant-attended treehoppers (Eurymelinae: Ipoini) have also been found in the region (Fletcher pers. comm.): Ipelloides macleayi from Tailem Bend – this species seems to be endemic to South Australia; Ipoides laeta from Lyrup Flats Reserve, west of Renmark, also been found in western NSW, western Qld and NT; and, Anacornutipoa lignosa from Waikerie – also found in WA and eastern Australia. Two species of treehoppers which live in ants' nests (Eurymelinae: Pogonoscopini) are also known from the region. Of the Membracidae (tree hoppers), a number of species, including Sextius spp., are associated with Acacia. Rigula calperum and Undarella pulleni were both described from Calperum Station.

Among the Suborder Heteroptera from the region is the widespread Pacific damsel bug (Nabis kinbergii) (Nabidae). Eggs are oviposited singly into grass stems and nymphs and adults feed on soft bodied insects, moth eggs, small larvae and mites.

The adults and nymphs of assassin bugs (Reduviidae) are predacious, almost entirely on other arthropods. They are usually solitary insects and are often conspicuous, sitting near the tips of shoots waiting for other insects to come within reach. Most are regarded as beneficial insects, for example

Coranus trabeatus, a ferocious reduviid which feeds on a wide range of insect pests. Larger species are capable of inflicting a painful 'bite' and may eject a harmful defensive fluid if handled.

The Lygaeidae (cinch bugs) is a large family with most species being seed-feeders. They are usually brown to red with a hard, elongated body ranging from 4-20 mm in length. The coon bug (Oxycarenus arctatus), in which the nymphs are small (~3 mm long), black and with blood-red abdomens, is common, and can be locally abundant on fences, at the base of walls and on areas such as ovals.

Coreidae are large bugs with strong repellent odours. Several species are found in the region, including the crusader bug on Acacia, Senna and Eucalyptus.

Nymph of Poecilometis sp.

Many different shield or stink bugs (Pentatomoidae) occur in the region. Scutelleridae include Austrotichus rugosus, which has been found in leaf litter at Keith, and Solenotichus circuliferus from Morgan and Loxton . A number of Poecilometis spp., often found under peeling bark of Eucalyptus or on Callitris, also occur. Poecilometis alienus has been collected from Morgan, Tailem Bend, and Karoonda to Peebinga, P. stigatus from along the Murray River, P. parilis from Sinclair Flat, upstream of Lock 1, and P. vermiculatus from Mercunda and Mannum. Several other genera occur in the region, often associated with the common trees of the region such as Eucualyptus, Callitris and Allocasuarina/Casuarina.

Coleoptera (Beetles)

Nearly all members of the Suborder Adephaga are nocturnal predators or scavengers, and several significant families occur in the region. Amongst the predatory Carabidae is the green tiger beetle (Megacephala australis), which can be locally abundant and hunts prey at dawn and dusk (crepuscular) on the banks of the Murray River. Also common in the region is the iridescent green carab beetle (Calosoma schayeri), a large, active species which hunts lepidopteran caterpillars and other slow-moving prey. It flies at night and is attracted to lights. These beetles can give off an unpleasant smell if handled. Fox scats are sometimes composed almost entirely of the remains of these beetles. Many other carabids (ground beetles) can be found, particularly in mallee vegetation. Also, adults of the aquatic families Dytiscidae (predaceous water beetles) and Gyrinidae (whirligig beetles) are often attracted to lights at night. Watts provided an identification guide to the genera of Australian water beetles.

About 90% of extant beetles belong to the Suborder Polyphaga, and these proportions seem to be exhibited in the beetle fauna of the Riverland and Murraylands. Several families of Staphylinoidea are found in the region, especially Staphylinidae (rove beetles). They are mostly predators or carrion feeders living in soil and leaf litter. Most are narrow elongate beetles with the posterior half of their abdomen exposed. As well, the carrion beetle (Ptomaphila lacrymosa) (Silphidae), is also common, and is most active in autumn and winter months.

Of the Scarabaeoidea, the families Trogidae, Lucanidae and Scarabaeidae are the most commonly encountered. Several species of Trox (Trogidae) are known from the region. Larvae burrow in the vicinity of carcasses, and both larvae and adults feed mainly on skin, fur and feathers. The larvae of lucanids (stag beetles) feed on dead wood, with the golden stag beetle (Lamprima aurata) being occasionally collected from Leptospermum, Acacia, Eucalyptus and sometimes from exotic tree species. Scarabs, however, are far more common and abundant than the other families. Geotrupinae are commonly seen in sandy areas after rain, especially Blackburnium spp. They feed on buried humus and underground fungi, and adults stridulate when caught. Aphodiinae, including the pasture cockchafer (Aphodius tasmaniae), are mostly dung-feeding beetles. The five horned dung beetle (Onthophagus pentacanthus) is common throughout much of the Murray-Darling Basin and further west in South Australia. Adults construct burrows for their young and provision them with animal dung. Adults are nocturnal and often attracted to lights. Several dung beetles which were introduced to control mainly cattle dung also occur in the region. Melolonthinae (chafers) have larvae which occur in the soil and feed on roots and humus, whilst adults are mostly short-lived and either do not feed at all, or feed on foliage. Of the many genera occurring in the region, the speciose Colpochila, Heteronyx and Liparetrus are associated with Eucalyptus. Christmas beetles (Anoplognathus spp.) can be seen flying in summer. Adults feed on Eucalyptus leaves and the larvae feed on rotting wood, humus or grass roots.

Buprestids or jewel beetles (Buprestidae) are widespread throughout the region. Some of the more colourful jewel beetles include Diadoxus, whose larvae bore into Callitris, Julidomorpha bakewelli which can occasionally be encountered in late December in the mallee, and Castiarina spp. which are mostly collected on flowers of plants such as Leptospermum.

Clerids (Cleridae), many of which are elongate, hairy beetles with distinctive colour patterns, are predacious as both larvae and adults. Depending on the species, these beetles can often be collected from flowers (especially Eucalyptus, Bursaria and Leptospermum), or by beating or sweeping vegetation.

Coccinellidae (ladybird beetles) are easily recognised due to their characteristic oval shape and usually bright, often spotted colour pattern. They range in size from 1 to 15 mm in length and have short antennae and legs, which are usually hidden beneath the body when viewed from above. The larvae are soft-bodied, and are often variously coloured with spots, and are usually adorned with spines. Almost all ladybirds are predatory as both larvae and adults, and feed on a variety of small soft-bodied arthropods such as aphids, scales and mites. As such, most ladybirds are viewed as beneficial insects. Apart from Rodolia cardinalis, important ladybirds in the region include the transverse ladybird (Coccinella transversalis) and common spotted ladybird (Harmonia conformis), with adults and larvae of both species being predators of soft-bodied insects. The mite-eating ladybird (Stethorus spp.) are also common but often overlooked because of its very small size. In contrast, both larvae and adults of the twentyeight-spotted potato ladybird (Henosepilachna vigintisexpunctata) feed on plants such as potato.

The Tenebrionoidea comprises some 28 families occurring in South Australia (Matthews 1987). They are predominantly associated with dead plant matter and the fungi that grow on it, although larvae of Rhipiphoridae and Meloidae are parasitic on other insects. Adults of several families (e.g. Mordellidae and Meloidae) are often collected from flowers. Of the Tenebrionidae, Lagria grandis is a common and widespread species with adults found feeding on Eucalyptus leaves. Celibe spp. and Helea spp. (pie-dish beetles) are found in leaf litter, especially in the mallee. Chalcopteroides spp. (darkling beetles) are conspicuous blue, green or purple metallic tenebrionids, mainly associated with eucalypts.

The Chrysomeloidea comprises two families, Cerambycidae and Chrysomelidae, both of which are speciose in the Riverland and Murraylands. Numerous longicorn beetles (Cerambycidae) are common, for example, Phoracantha spp. which produce oval emergence holes in eucalypts such as the sugar gum. Adults of a number of genera are often seen at lights. These include Eurynassa spp. whose larvae bore into the dead wood of Acacia. Matthews has an extensive list of longicorn beetle genera and their host plants in South Australia. Most Chrysomelidae (leaf beetles) are leaf-feeding as both larvae and adults, although larvae may exhibit other habits such as stem or root boring and seed feeding. Matthews and Reid list chrysomelid beetle genera and their host plants in the state. Amongst the common species are Paropsis and Paropsisterna which feed mostly on eucalypts.

Many species of weevils (Curculionidae) are found in the region. Common weevils include the large and spectacular diamond beetle (Chrysolopus spectabilis) which feeds on various acacias at locations such as Monarto South, elephant weevil (Orthorhinus cylindrirostris), Eucalyptus snout weevils (Gonipterus spp.), and several species of curculio beetles (Otiorhynchus), some of which are considered pest species. Amongst the smallest weevils are Misophrice spp. which are about 2 mm long and found on Allocasuarina/Casuarina where they are believed to be pollen feeders.

Diptera (Mosquitoes, Midges and Flies)

Many other families and species are found in the region. Flies are distinguished by having a single pair of wings on the thorax and a pair of halteres or balancing organs (tiny reduced hind wings).

Tabanidae (March flies and horse flies) are medium to large biting flies, and several species are known from the region. Larvae are carnivorous, preying on insects and snails, and may be found living in damp soil or rotting logs, as well as in aquatic environments. Adults of most Australian species are blood-suckers, but some are flower-feeders (e.g. Scaptia). The blood-suckers are voracious feeders, often hovering before landing and piercing the skin with their needle-like proboscis and absorbing a considerable amount of blood.

Robber flies (Asilidae) are a widely distributed group of predatory flies which are diverse in semi-arid and arid regions. Adult robber flies prey on insects, although some capture spiders. Several species of Bathypogon, Cerdistus, Colepia and Mauropteron are known from the region.

Beeflies (Bombyliidae) and hover flies (Syrphidae) are common and widespread in the region. Amongst the more colourful beeflies (Exoprosopini) are Balaana centrosa, Kapua adelaidica and K. irwini collected near Calperum Station, which have distinct clear and black wing patterns . Hover flies of the region include members of the common genera Simosyrphus and Eristalis, the larvae of which prey on soft-bodied insects.

Tachinid flies (Tachinidae) can be small to large with wingspans up to 35 mm depending on the species. Most have stout bodies covered in bristles and are either grey or black, although some have a metallic appearance and, as a consequence, superficially resemble blowflies. The larvae of tachinids are parasitoids of other insects.

Lepidoptera (Moths and Butterflies)

Many of the moths listed for the Adelaide Region in McQuillan and Forrest (1985) occur in the Riverland and Murraylands, and here are highlighted just some of the more common species. In autumn, various ghost moths (Hepialidae) can be seen, including the Bardi (Barti) moth (Trictena argentata), which is common along the banks of the Murray River at Waikerie when they emerge from their pupal cases in the ground near trees. Larvae, pupae and adults were eaten by Aboriginals.

The giant wood moth (Xyleutes liturata) (Cossidae), is the largest moth in South Australia with a wingspan of up to 200 mm in females and 130 mm in males. Its larvae bore into the woody tissue of eucalypts, particularly the river red gum. Several species of case moth (Psychidae) occur in the region, including the faggot case moth (Clania ignobilis) whose older larvae form distinctive larval cases surrounded by twigs of their eucalypt food-plant.

Trictena argentata.

An interesting moth occurring in the region is Stathmopoda melanochra (Stathmopodidae), which deposits eggs amongst colonies of gum tree scales (e.g. Eriococcus coriaceus) where the larvae devour them. A number of geometrid moths are common including the red-lined geometrid (Crypsiophona ocultaria), angled satin moth (Thalaina angulosa) and hakea moth (Oenochroma vinaria). The gumleaf skeletonizer (Uraba lugens), often found on river red gum, is interesting in that the caterpillars have a characteristic 'spike' on top of their head, comprising old head capsules which are shed as they grow. One noctuid, the granny moth (Dasypodia selenophora), whose larvae feed on Acacia, is commonly seen in late spring resting during daylight hours under eaves or on ceilings.

Butterflies can generally be distinguished from moths by a number of characters, including their day-flying habits, bright colours, presence of a clubbed antenna and the ability to close their wings dorsally over the thorax. However, there are exceptions.

Butterflies belong to two superfamilies, Hesperioidae and Papilionoidea (several families), with most Australian species belonging to the latter. Braby covers the described butterflies of the region, providing descriptions, habits and larval food plants. As well, South Australian butterflies are extensively treated by Grund. Several skippers (Hesperidae) can be found in the Riverland and Murraylands, including the sciron ochre or mallee ochre (Trapezites sciron), whose males can be observed hilltopping on sand dune ridges. In addition to the small citrus butterfly (P. anactus) and occasionally the large citrus butterfly (P. aegus aegus), the only other Papilionidae seen in the region is the chequered swallowtail (P. demoleus). Pieridae (whites and yellows) include the cabbage white butterfly, the white migrant (Catopsilia pyranthe), which occasionally migrates into the region but does not breed, the common small grass yellow (Eurema smilax), and the spotted jezebel (Delias aganippe).

Nymphalidae are medium to large butterflies and are commonly brown or orange. The common brown (Heteronympha merope) is widespread in the region and its larval food plants include Poa spp. and kangaroo grass (Themeda trianda). The meadow argus (Junonia villida), Australian painted lady (Vanessa kershawi), yellow admiral (V. itea), lesser admiral (Danaus chrysippus), and monarch or wanderer butterfly (D. plexippus), are all widespread and common.

(left), small grass yellow, Eurema smilax; 15 (right), inland form of the amaryllis azure (Ogyris amaryllis meridionalis,) female.

Lycaenidae (blues and azures) are small to medium-sized butterflies, often with wings which have an iridescent blue, purple or green upper and brown or white patterned underside of their wings. Numerous species occur in the region. Perhaps the most often encountered is the common grass-blue (Zizina labradus) which occurs throughout Australia. The larvae eat the flowers, leaves and immature seeds within developing seed pods of a wide variety of native and introduced legumes, including lucerne. Other species include the amaryllis azure (Ogyris amaryllis meridionalis), whose larvae feed on various mistletoes (Amyema spp.) growing on Acacia and Casuarina/Allocasuarina. They can be attended by a range of ant species, including Iridomyrmex and Camponotus. The amethyst hairstreak (Jalmenus icilius) also occurs in the region. Its larvae feed on Acacia and Senna and are attended by Iridomyrmex. Other blues include the two-spotted lined blue (Nacaduba biocellata) whose larvae live on flower buds and flowers of Acacia, wattle blue (Theclinesthes miskini) whose larvae feed mainly on Acacia, saltbush blue (T. serpentata) whose larvae feed on chenopods such as Atriplex and Rhagodia, and the long-tailed pea-blue (Lampides boeticus) whose larvae feed on Fabaceae.

Hymenoptera (Sawflies, Wasps, Ants and Bees)

As with other megadiverse orders (Coleoptera, Diptera Lepidoptera and Hemiptera), several thousand species of Hymenoptera occur in the region. Many of the parasitic wasps are 2 mm or less in length, and are difficult to observe and identify even to family level without the aid of a stereo-microscope and specialist taxonomic keys. The order is divided traditionally into Symphyta and Apocrita, with the former being the 'primitive' sawflies. Apocritans include wasps, bees and ants, and are distinguished from the Symphyta by the narrow 'waist' (petiole) formed between the first two segments of the actual abdomen.

Of the sawfly families (symphytans), Pergidae is the most common, with members of the genus Perga being the most abundant. Their larvae or 'spitfires', so called because the aggregating larvae rear up when disturbed and regurgitate eucalypt oils, are common on Eucalyptus. Adult pergids are, however, rarely seen. Leaf blister sawflies (Phylacteophaga) are also common, and mine the young leaves of river red gum and other eucalypts.

Among the apocritan wasps, two families belonging to the Superfamily Evanioidea (Evaniidae and Gasteruptiidae) are widespread in the Riverland and Murraylands. Evaniidae (hatchet or ensign wasps) parasitise cockroach oothecae and adults of several species of Acanthinevania and Szepligetella can be collected on various flowers, including eucalypts, Melaleuca and Bursaria, as adults use nectar as an energy source. The larvae of Gasteruptiidae occur in the nests of solitary bees and wasps where they kill the host egg or larvae and then consume the pollen store provided by the host for its developing larvae (i.e. predator-inquilines). As with evaniid wasps, adult gasteruptiids can be found on flowers with an abundant nectar supply or can be seen searching for host nests on tree trunks, banks or ground. One gasteruptiid, Pseudofoenus morganensis, has only been collected near Morgan.

(Left) 'Spitfires' of Perga sp., (Right) Gasteruption sp., female.

Ichneumonid wasps (Ichneumonidae) can be small to very large and are commonly seen at flowers although some species are attracted to light, including Enicospilus spp., which are ectoparasitoids of large lepidopteran larvae or symphytan wasps. Whilst many ichneumonids parasitise lepidopteran larvae, others parasitise Hymenoptera, Neuroptera and Coleoptera. Braconids (Braconidae) are minute to large wasps, and amongst the genera in the region are Apantales, Chelonus, Cotesia, and Bassus (parasitoids of moth larvae), Jarra (parasitoids of beetle larvae), Aspilota and Opius (parasitoids of dipteran larvae) and Aphidius (parasitoids of aphids).

Platygastrid wasps (Platygastroidea) include species which parasitise Cecidomyiidae (Diptera) leaf and stem galls on various plants including Eucalyptus and Acacia. Amongst the Scelionidae, is the genus Scelio, which are parasitoids of locust and grasshopper eggs. Several species are known from the region, including S. fulgidus (Murray Bridge - bred from eggs of the Australian plague locust), S. fulvithorax (Renmark and Bookmark Biosphere Reserve), S. imporcerus (near Keith), and S. parvicornis (Waikerie).

Many of the ~19 Australian families classified in the Superfamily Chalicidoidea can be found in the region. They are mostly small to minute parasitic wasps. Of these families, species of Pteromalidae have been reared from Apiomorpha galls (Hemiptera), Aphelinidae including Aphytis (parasitoids of diaspid scales including red scale on citrus) and Aphelinus (aphid parasitoids), and Trichogrammatidae (Trichogramma spp. which parasitise moth eggs, including Helicoverpa) are common.

Several chrysidid or cuckoo wasps (Chrysididae) occur in the region. The large metallic blue Stilbum cyanurum parasitises mud dauber wasps (Sceliphron) and potter wasps (Abispa) and has probably been introduced.

(Left) Cuckoo wasp (Chrysididae), (Right) Polistes nest.

The Superfamily Vespoidea (7 families) is very well represented. Pompilidae (spider wasps) are small to large predatory wasps that all hunt spiders or, more rarely, other pompiliids. Several velvet ant species (Mutillidae) occur in the region. The wingless females, which superficially resemble large hairy ants, run around on the ground, tree trunks or walls searching for nests of Vespidae and Apidae. Female velvet ants are capable of delivering a painful sting if disturbed. Numerous species of flower wasps (Tiphiidae) also occur in the region. Most are believed to be parasitic on soil-dwelling beetle larvae, and all species have wingless females. In these species, the male often carries the female from flower to flower while mating. One obvious species is the 'blue ant' (Diamma bicolor) which parasitises mole crickets. The female is about 23 mm long, has a polished blue-green body and reddish legs, and can deliver a painful sting if disturbed. Some species of flower wasps are also implicated in the pollination of sexually deceptive orchids such as the endangered coloured spider orchid, Caladenia colorata. Vespid wasps (Vespidae) are commonly encountered, including potter wasps (Abispa) and paper wasps (Polistes). Polistes nests are common under ledges on the cliffs along the Murray River (e.g. at Blanchetown and Waikerie) and under eaves of buildings.

Many genera and species of ants (Formicidae) are represented. Whilst the majority of ants are general predators or scavengers, feeding on a wide range of prey including other arthropods, many are seed feeders. Some also tend sap-sucking Hemiptera and other insects. Among the commonly encountered ants in the region are the meat ants (Iridomyrmex spp.), which have nests that can be of an enormous size with tens of thousands of workers. They clear vegetation from the surface around their nest and cover the area with small stones. Some Iridomyrmex are associated with butterflies, including the satin azure (Ogyris amaryllis). Another common genus is Camponotus (sugar ants). They can be seen foraging on the ground and on vegetation by day but more frequently by night. Adults derive their sustenance from the excreta of sap-sucking insects, floral nectaries and seeds. Several species are known to have a mutualistic relationship with Ogyris butterflies. Other common ant genera in the region include Melophorus, Meranoplus (especially in the mallee), Myrmecia, Pheidole and Rhytidoponera.

(left) inch ant (Myrmecia sp.); 21 (right) male bluebanded bee (Amegilla sp.).

Apoidea from the region include Sphex (Sphecidae) which provisions its underground nest with crickets (Gryllidae) and katydids (Tettigonidae), the common mud dauber (Sceliphron laetum) (Sphecidae) which provisions its mud nest with spiders, and Bembix (Crabrionidae) which provisions its underground nests with flies or occasionally bees or other wasps.

Bees all belong to the Family Apidae, and all native bees in the region are solitary (not social as is the case with the honey bee). They are commonly seen collecting nectar and/or pollen on a wide range of flowers, and include the genera Ctenocolletes, Euryglossa, Hylaeus, Leioproctus and Nomia. Some of the more conspicuous bees are the solitary blue-banded bees (Amegilla spp.), which have pale blue bands on a black abdomen. At least one species of these bees shows potential as a pollinator of glasshouse grown tomatoes.

Permissions

All chapters in this book are published with permission under the Creative Commons Attribution Share Alike License or equivalent. Every chapter published in this book has been scrutinized by our experts. Their significance has been extensively debated. The topics covered herein carry significant information for a comprehensive understanding. They may even be implemented as practical applications or may be referred to as a beginning point for further studies.

We would like to thank the editorial team for lending their expertise to make the book truly unique. They have played a crucial role in the development of this book. Without their invaluable contributions this book wouldn't have been possible. They have made vital efforts to compile up to date information on the varied aspects of this subject to make this book a valuable addition to the collection of many professionals and students.

This book was conceptualized with the vision of imparting up-to-date and integrated information in this field. To ensure the same, a matchless editorial board was set up. Every individual on the board went through rigorous rounds of assessment to prove their worth. After which they invested a large part of their time researching and compiling the most relevant data for our readers.

The editorial board has been involved in producing this book since its inception. They have spent rigorous hours researching and exploring the diverse topics which have resulted in the successful publishing of this book. They have passed on their knowledge of decades through this book. To expedite this challenging task, the publisher supported the team at every step. A small team of assistant editors was also appointed to further simplify the editing procedure and attain best results for the readers.

Apart from the editorial board, the designing team has also invested a significant amount of their time in understanding the subject and creating the most relevant covers. They scrutinized every image to scout for the most suitable representation of the subject and create an appropriate cover for the book.

The publishing team has been an ardent support to the editorial, designing and production team. Their endless efforts to recruit the best for this project, has resulted in the accomplishment of this book. They are a veteran in the field of academics and their pool of knowledge is as vast as their experience in printing. Their expertise and guidance has proved useful at every step. Their uncompromising quality standards have made this book an exceptional effort. Their encouragement from time to time has been an inspiration for everyone.

The publisher and the editorial board hope that this book will prove to be a valuable piece of knowledge for students, practitioners and scholars across the globe.

Index